Neuromorphic Computing and Beyond

Khaled Salah Mohamed

Neuromorphic Computing and Beyond

Parallel, Approximation, Near Memory, and Quantum

 Springer

Khaled Salah Mohamed
A Siemens Business
Mentor
Heliopolis, Egypt

ISBN 978-3-030-37226-2   ISBN 978-3-030-37224-8   (eBook)
https://doi.org/10.1007/978-3-030-37224-8

This Springer imprint is published by the registered company Springer Nature Switzerland AG
The registered company address is: Gewerbestrasse 11, 6330 Cham, Switzerland

Dedicated to the memory of my father who inspired me.

Preface

This book brings together the fields of neuromorphic, approximate, parallel, in-memory, and quantum computing. All these paradigms are used to enhance computing capability as Moore's law is slowing down, but the demand for computing keeps increasing. This book provides an elementary introduction for students and researchers interested in these approaches and gives a state-of-the-art overview and the essential details of these paradigms. Moreover, it provides a good introduction and comprehensive discussions into the neuromorphic, approximate, parallel, in-memory, and quantum computing concepts. This book will be useful to anyone who needs to rapidly gain a broad understanding of neuromorphic, approximate, parallel, in-memory, and quantum computing. This book brings the reader to the most advanced theories develop thus far in this emerging research area.

- Neuromorphic computing: It has become one of the most important computationally intensive applications for a wide variety of fields such as image or object recognition, speech recognition, and machine language translation.
- Parallel computing: A performance comparison of MPI, Open MP, and CUDA parallel programming languages are presented. The performance is analyzed using cryptography as a case study.
- Quantum computing: It uses quantum phenomenon to perform computational operations. Operations are done at an atomic level. What can be done with single qubits is very limited. For more ambitious information processing and communication protocols, multiple bits must be used and manipulated. It can be used in many complex applications such as speed-up prime factoring and sorting huge database.
- Near-memory and in-memory computing: They solve the bandwidth bottleneck issues in today's systems as they enable moving from big data era to fast data era. They reduce data movement by performing processing-in-memory or near-memory.
- Approximate computing: Sacrificing exact calculations to improve performance in terms of run-time, power, and area is at the foundation of approximate computing.

This book shows that the approximate computing is a promising paradigm towards implementing ultra-low power systems with an acceptable quality for applications that do not require exact results.

Heliopolis, Egypt   Khaled Salah Mohamed

Contents

Chapter 1
An Introduction: New Trends in Computing

1.1 Introduction

The development of IC technology is driven by the needs to increase performance and functionality while reducing size, weight, power consumption, and manufacturing cost. As Gordon Moore predicted in his seminal paper, reducing the feature size also allows chip area to be decreased, improving production, and thereby reducing cost per function. The scaling laws showed that improved device and ultimately processor speed could be achieved through dimensional scaling. However, all trends ultimately have limits, and Moore's law is no exception. The limits to Moore's law scaling have come simultaneously from many directions. Lithographic limits have made it extremely difficult to pack more features onto a semiconductor chip, and the most advanced lithographic techniques needed to scale are becoming prohibitively expensive for most Fabs. The optical projection systems used today have very complex multielement lenses that correct for virtually all of the common aberrations and operate at the diffraction limit. The resolution of a lithography system is usually expressed in terms of its wavelength and numerical aperture (NA) as:

$$\text{Resolution} = k_1 \frac{\lambda}{\text{NA}} \qquad (1.1)$$

where k_1, the constant, is dependent on the process being used. In IC manufacturing, typical values of k_1 range from 0.5 to 0.8, with a higher number reflecting a less stringent process. The NA of optical lithography tools ranges from about 0.5 to 0.6 today. Thus, the typical rule of thumb is that the smallest features that can be printed are about equal to the wavelength of the light used. Historically, the improvements in IC lithography resolution have been driven by decreases in the printing wavelength. The illumination sources were initially based on mercury arc lamps filtered for different spectral lines. The figure shows the progression from G-line at 435 nm to I-line at 365 nm. This was followed by a switch to excimer laser sources with

© Springer Nature Switzerland AG 2020
K. S. Mohamed, *Neuromorphic Computing and Beyond*,
https://doi.org/10.1007/978-3-030-37224-8_1

KrF at 248 nm and, more recently, ArF at 193 nm. The most advanced IC manufacturing currently uses KrF technology with the introduction of ArF tools beginning sometime in 2001. It can also be seen from the figure that the progress in IC minimum feature size is on a much steeper slope than that of lithography wavelength. Prior to the introduction of KrF lithography, the minimum feature sizes printed in practice have been larger than the wavelength with the crossover at the 250-nm generation and KrF. With the introduction of 180-nm technology in 1999, the most advanced IC manufacturing was done with feature sizes significantly below the wavelength (248 nm).

Furthermore, short-channel effects and random fluctuations are making conventional planar device geometries obsolete. Interconnect becomes more significant limiting factor to power dissipation and performance of a chip. Wires get closer to each other and the length of interconnect increases as a result of larger die size as feature size decreases. Interconnect capacitance and resistance increase while device parasitic reduce as wires get closer to each other and wire thickness does not shrink in the same scale as device size reduces. Finally, the fact that scaling has proceeded without appreciable voltage reduction over the past decade has increased power densities to the precipice of cooling and reliability limits [1].

We need a dramatically new technology to overcome these CMOS limitations and offer new opportunity to achieve massive parallelism. Moreover, certain types of problems such as learning, pattern recognition, fault-tolerant system, cryptography, and large set search algorithms are intrinsically very difficult to solve even with fast evolution of CMOS technology. Fundamental limits on serial computing can be summarized as Three "Walls" limitations.

1.1.1   Power Wall

Increasingly, microprocessor performance is limited by achievable power dissipation rather than by the number of available integrated-circuit resources. Thus, the only way to significantly increase the performance of microprocessors is to improve power efficiency at about the same rate as the performance increase. This is due to the fact that power is increasing with frequency. An example is shown in Fig. 1.1.

Fig. 1.1 Power wall example

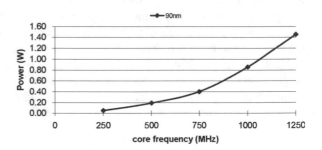

From the early 1990s to today, power consumption has transitioned into a primary design constraint for nearly all computer systems.

1.1.2 Frequency Wall

Conventional processors require increasingly deeper instruction pipelines to achieve higher operating frequencies. This technique has reached a point of diminishing returns, and even negative returns if power is taken into account. Processors are choosing to trade off performance by lowering supply voltage. The performance loss of reduced voltage and clock frequency is compensated by further increased parallelism, where the power is giving by:

$$P = f c_l v^2 \qquad (1.2)$$

where f is the frequency, c_l is the load capacitance, and v is the voltage.

1.1.3 Memory Wall

On multi-gigahertz symmetric processors, latency to DRAM memory is currently approaching 1000 cycles. As a result, program performance is dominated by the activity of moving data between main storage and the processor. In other words, memory technology has not been able to keep up with advancements in processor technology in terms of latency and energy consumption [2]. Moreover, there is a memory capacity problem where memory capacity per core is expected to drop by 30% every 2 years as shown in Fig. 1.2.

Designers have explored larger on-chip caches, faster on-chip interconnects, 3D integration, and a host of other circuit and architectural innovations to address the processor-memory gap. Yet, data-intensive workloads such as search, data analytics, and machine learning continue to pose increasing demands on on-chip memory systems, creating the need for new techniques to improve their energy efficiency and performance.

1.2 Classical Computing

Classical Computer is made up of main memory, arithmetic unit, and control unit. Transistor is the most basic component of computer. Transistors basically just work as switch. Transistor size is an important part of improving computer technology. Today's size of transistor is 14 nm. As transistor is getting smaller and smaller, a new problem arises. At that scale laws Quantum mechanics starts to influence and

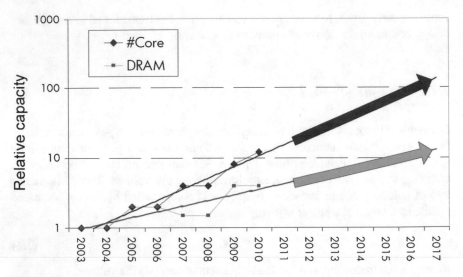

Fig. 1.2 The memory capacity gap

transistor cannot work properly due quantum tunneling. With N transistors, there are 2 N different possible states of the computer. There are three key components in any computing system: computation, communication, and storage/memory. Today's memory hierarchy usually consists of multiple levels of cache, a main memory, and storage [3–9]. Every computer needs an operating system (OS). It acts as a layer of communication between the user and the components. The choice of OS depends on:

- Software availability.
- Performance.
- Reliability.
- Hardware compatibility.
- Ease of administration/maintenance costs.

1.2.1 Classical Computing Generations

The first computers used vacuum tubes for circuitry and magnetic drums for memory. They were often enormous, taking up entire rooms, and their processing capabilities were very slow. They were very expensive to operate. Then transistors replaced vacuum tubes. One transistor replaced the equivalent of nearly 40 vacuum tubes. This allowed computers to become smaller, faster, cheaper, more energy-efficient, and more reliable. Programmers started to use "FORTRAN" and "COBOL" languages to operate computers. Afterwards, silicon-based integrated circuits were first used in building those computers where it increases the speed and

Table 1.1 Classical computing generations

Year	Generation	Type
1946–1958	First generation	Vacuum tubes
1959–1964	Second generation	Transistors
1965–1970	Third generation	Integrated circuits
1971–1980	Fourth generation	Microprocessors
1981	Fifth generation	VLSI: first personal computers (IBM)
1984	Fifth generation	First MAC computer (Apple)
1991	Fifth generation	The internet

efficiency of computers. Compared to the second-generation computers, computers became smaller and cheaper. The microprocessor brought the fourth generation of computers, as thousands of integrated circuits were built onto a single silicon chip. This enabled computers to become small and more powerful. The first Personal Computers (PC) was introduced by IBM Corp in 1981. In 1984, Apple Corp introduced the first Mac Computer. It was the first computer with Graphic User Interface (GUI) and with a mouse. In 1991, the first internet web page was built. Table 1.1 summarizes classical computing generations [10, 11].

1.2.2   Types of Computers

- Personal computers (PCs): Actual personal computers can be generally classified by size and chassis/case.
- Desktop computers: A workstation is a computer intended for individual use that is faster and more capable than a personal computer. It is intended for business or professional use (rather than home or recreational use).
- Notebook (laptop) computers: A small, portable computer—small enough that it can sit on your lap. Nowadays, laptop computers are more frequently called notebook computers.
- Handheld computers/Tablet PCs: A portable computer that is small enough to be held in one's hand.
- PDA (personal digital assistant): Short for personal digital assistant, a handheld device that combines computing, telephone/fax, and networking features.
- Mainframe computers: A mainframe is a high-performance computer used for large-scale computing purposes that require greater availability and security than a smaller-scale machine can offer.
- Supercomputers: A supercomputer is a very powerful machine, mainly used for performing tasks involving intense numerical calculations such as weather forecasting, fluid dynamics, nuclear simulations, theoretical astrophysics, and complex scientific computations.

1.3   Computers Architectures

Most modern CPUs are heavily influenced by the Reduced Instruction Set Computer (RISC) design style. With RISC, the focus is to define simple instructions such as load, store, add, and multiply. These instructions are commonly used by the majority of applications and then to execute those instructions as fast as possible. Also, there is Complex Instruction Set Computer (CISC), where we reduce the number of instructions per program ("LOAD" and "STORE" are incorporated in instructions). CPU performance is determined by the below equation [12].

$$\text{CPU performance} = \frac{\text{Time}}{\text{Program}} = \frac{\text{Time}}{\text{cycle}} \times \frac{\text{cycles}}{\text{instruction}} \times \frac{\text{instructions}}{\text{program}} \qquad (1.3)$$

1.3.1   Instruction Set Architecture (ISA)

There are no relations between Instruction Set (**RISC and CISC**) with architecture of the processor (Harvard Architecture and Von-Neumann Architecture). Both instruction sets can be used with any of the architecture. Examples of and differences between CISC and RISC are shown in Table 1.2. Very long instruction word

Table 1.2 RISC and CISC examples and characteristics

RISC example	CISC example
Load A, 1000 Load B, 1001 Mul A, B Store 1002, A	Mul A, B
Characteristics of RISC architecture	**Characteristics of CISC architecture**
• Simple instructions are used in RISC architecture. • RISC helps and supports few simple data types and complex data types. • RISC utilizes simple addressing modes and fixed length instructions for pipelining. • One cycle execution time. • The amount of work that a computer can perform is reduced by separating "LOAD" and "STORE" instructions. • In RISC, pipelining is easy as the execution of all instructions will be done in a uniform interval of time, i.e., one click. • In RISC, more RAM is required to store assembly-level instructions. • Reduced instructions need a smaller number of transistors in RISC.	• Instruction-decoding logic will be complex. • One instruction is required to support multiple addressing modes. • Less chip space is enough for general purpose registers for the instructions that are operated directly on memory. • Various CISC designs are set up two special registers for the stack pointer, handling interrupts, etc. • MUL is referred to as a "complex instruction" and requires the programmer for storing functions.

(**VLIW**) is another type of instruction set which refers to instruction set architectures designed to exploit instruction-level parallelism (ILP).

An ISA may be classified by architectural complexity. A complex instruction set computer (CISC) has many specialized instructions, some of which may be used in very specific programs. A reduced instruction set computer (RISC) simplifies the processor by implementing only the instructions that are frequently used in programs, while the fewer common operations are implemented as subroutines. ISA's instructions can be categorized by their type:

* Arithmetic and logic operations.
* Data handling and memory operations.
* Control flow operations.
* Coprocessor instructions.
* Complex instructions.

An instruction consists of several operands depending on the ISA; operands may identify the logical operation and may also include source and destination addresses and constant values. On traditional architectures, an instruction includes an opcode that specifies the operation to perform. Many processors have fixed instruction widths but have several instruction formats. The actual bits stored in a special fixed-location "instruction type" field (that is in the same place in every instruction for that CPU) indicates which of those instruction formats is used by this specific instruction—which particular field layout is used by this instruction. For example, the **MIPS processors** have R-type, I-type, J-type, FR-type, and FI-type instruction formats. The size or length of an instruction varies widely depending on the ISA; within an instruction set, different instructions may have different lengths. A RISC instruction set normally has a fixed instruction length (often 4 bytes = 32 bits), whereas a typical CISC instruction set may have instructions of widely varying length.

RISC-V is an open and free Instruction Set Architecture (ISA). The ISA consists of a mandatory base integer instruction set (denoted as RV32I, RV64I, or RV128I with corresponding register widths) and various optional extensions denoted as single letters, e.g., M (integer multiplication and division) and C (compressed instructions). Thus, RV32IMC denotes a 32 bit core with M and C extension [13]. The instruction set is very compact, RV32I consists of 47 instructions and the M extension adds additional 8 instructions. All RV32IM instructions have a 32 bit width and use at most two source and one destination register. The C extension adds 16 bit encodings for common operations. The RISC-V ISA also defines Control and Status Registers (CSRs), which are registers serving a special purpose. Furthermore, the ISA provides a small set of instructions for interrupt handling and interacting with the system environment [14].

SPARC (Scalable Processor Architecture) is a reduced instruction set computing instruction set architecture originally developed by Sun Microsystems and Fujitsu [15].

1.3.2 Different Computer Architecture

1.3.2.1 Von-Neumann Architecture: General-Purpose Processors

CPUs are designed to run almost any calculation; they are general-purpose computers. To implement this generality, CPUs store values in registers, and a program tells the Arithmetic Logic Units (ALUs) which registers to read, the operation to and the register into which to put the result. A program consists of a sequence of these read/operate/write operations. Von-Neumann architecture consists of memory (RAM), central processing unit (CPU), control unit, arithmetic logic unit (ALU), and input/output system (Fig. 1.3). Memory stores **program and data**. Program instructions execute sequentially. These computers employ a fetch-decode-execute cycle to run programs. The control unit fetches the next instruction from memory using the program counter to determine where the instruction is located. The instruction is decoded into a language that the ALU can understand. Any data operands required to execute the instruction are fetched from memory and placed into registers within the CPU. The ALU executes the instruction and places results in registers or memory. The operation can be summarized in the following steps (Fig. 1.4) [16]:

1. Instruction fetch: The value of PC is outputted on address bus, memory puts the corresponding instruction on data bus, where it is stored in the IR.
2. Instruction decode: The stored instruction is decoded to send control signals to ALU which increment the value of PC after pushing its value to the address bus.
3. Operand fetch: The IR provides the address of data where the memory outputs it to ACC or ALU.
4. Execute instruction: ALU is performing the processing and store the results in the ACC. The processors can be programed using high-level language such as C or mid-level language such as assembly. Assembly is used, for example, in nuclear application because it is more accurate. At the end the compiler

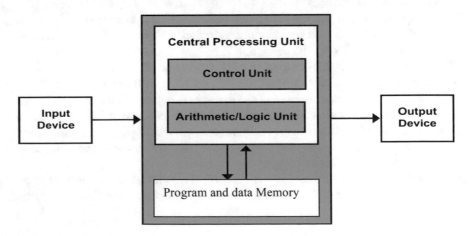

Fig. 1.3 Von-Neumann architecture

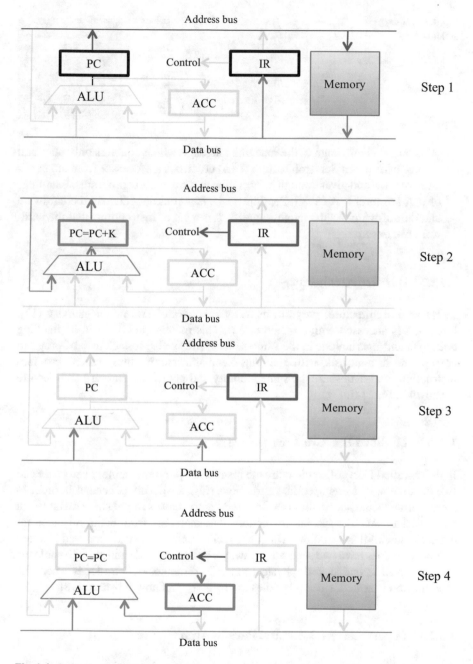

Fig. 1.4 A simple processor operation

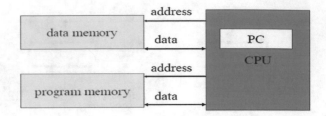

Fig. 1.5 Harvard
architecture

translates this language to the machine language which contains only ones and
zeroes. Instruction Set Architecture (ISA) describes a processor from the user's
point of view and gives enough information to write correct programs. Examples
of ISA are Intel ISA (8086, Pentium). ISA is a contract between the hardware
and the software. As the name suggests, it is a set of instructions that the hardware can execute.

1.3.2.2   Harvard Architecture

In Harvard architecture, program memory is separated from data memory [17].
Each part is accessed with a different bus. This means the CPU can be fetching
both data and instructions at the same time. There is also less chance of program
corruption. It contrasts with the Von-Neumann architecture, where program
instructions and data share the same memory and pathways. Harvard architecture
is shown in Fig. 1.5.

1.3.2.3   Modified Harvard Architecture

In the modified Harvard architecture the instruction/program memory can be treated
like data memory using specific instructions (Fig. 1.6). This is needed in order to
store constants and access them easily. Modern processors might share memory but
have mechanisms like special instructions that keep data from being mistaken for
code. Some call this "modified Harvard architecture." However, modified Harvard
architecture does have two separate pathways: busses for code and storage while the
memory itself is one shared, physical piece. The memory controller is where the
modification is seated, since it handles the memory and how it is used [18].

1.3.2.4   Superscalar Architecture: Parallel Architecture

A superscalar processor is a CPU that implements a form of parallelism called
instruction-level parallelism within a single processor. In contrast to a scalar processor that can execute at most one single instruction per clock cycle, a superscalar processor can execute more than one instruction during a clock cycle by simultaneously

Fig. 1.6 Modified Harvard
architecture

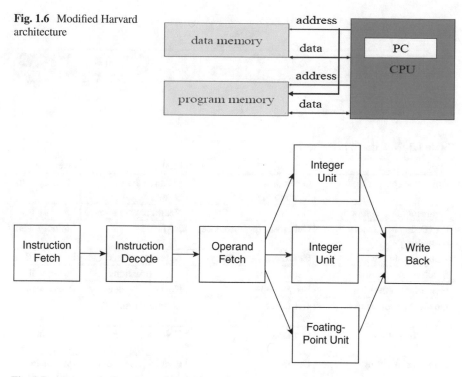

Fig. 1.7 An example for superscalar architecture

dispatching multiple instructions to different execution units on the processor. It
therefore allows for more throughput (the number of instructions that can be executed
in a unit of time) than would otherwise be possible at a given clock rate. Each execution unit is not a separate processor (or a core if the processor is a multicore processor), but an execution resource within a single CPU such as an arithmetic logic unit.
An example for superscalar architecture with three functional unit is shown in Fig. 1.7.

1.3.2.5 VLIW Architecture: Parallel Architecture

Superscalar processor implements a form of parallelism by executing more than
one instruction simultaneously based on some hardware-based scheduling techniques. VLIW executes operations in parallel based on a fixed schedule determined
during the compile time. VLIW approach makes extensive use of the compiler by
requiring it to incorporate several small independent operations into a long instruction word. The instruction is large enough to provide, in parallel, enough control
bits over many functional units. In other words, a VLIW architecture provides
many more functional units than a typical processor design, together with a compiler that finds parallelism across basic operations to keep the functional units as
busy as possible. The compiler compacts ordinary sequential codes into long

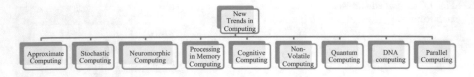

Fig. 1.8 New trends in computing

Table 1.3 New trends in computing

Type	Aim	Methods	Overcomes
Cognitive computing	Get smarter (decision-making)	Machine learning based	Memory wall
Neuromorphic computing	Get smarter	Neural network based	Memory wall
Processing in memory computing	Get closer (performance)	3D based	Memory wall
Approximate computing	Get tolerant (accuracy)	Approximations	Power wall
Stochastic computing	Get tolerant	Approximations	Power wall
Nonvolatile computing	Error-resilient	Nonvolatile memory	Power wall
Quantum computing	Get faster	Quantum principles	Frequency wall
DNA computing	Get faster	DNA principles	Frequency wall
Parallel computing	Get faster	Parallel programming	Frequency wall

instruction words that make better use of resources. During execution, the control unit issues one long instruction per cycle. The issued instruction initiates many independent operations simultaneously.

1.4   New Trends in Computing

For decades, enhancing computing capability is driven by the scaling of CMOS transistors according to Moore's law. However, both the CMOS scaling and the classical computer architecture are approaching fundamental and practical limits as discussed previously. The size of transistors has been scaled down to a few nanometers and further shrinking will eventually reach the atomic scale. Therefore, new computing trends and architectures are needed. Figure 1.8 summarizes the new computing trends that we will discuss through this book. Table 1.3 lists the aim of each trends and how we achieve it. We are aiming at making computers faster, cheaper, smaller, and more reliable. All these trends will be discussed thoughtfully in the remaining chapters. The proposed techniques intend to improve power and performance of computing without relying on technology shrinking.

1.5 Conclusions

In this chapter, we introduce the traditional computing concepts. We highlight the main walls and the need for new trends in computing to enhance computing capability as Moore's law slowing down, but the demand for compute keeps increasing. Many trends are introduced and it will be discussed thoroughly in the next chapters.

References

1. A. Moshovos, *Advanced Computer Architecture* (U. of Toronto, Toronto, 2018)
2. W.A. Wulf, S.A. McKee, Hitting the memory wall: Implications of the obvious. SIGARCH Comput. Archit. News **23**(1), 20–24 (1995)
3. https://blogs.nvidia.com/blog/2009/12/16/whats-the-difference-between-a-cpu-and-a-gpu/
4. https://www.maketecheasier.com/difference-between-cpu-and-gpu/
5. https://en.wikipedia.org/wiki/Graphics_processing_unit
6. https://www.youtube.com/watch?v=lGefnd7Fmmo (Nvidia GPU Architecture)
7. https://en.wikipedia.org/wiki/CUDA
8. https://www.anandtech.com/show/8526/nvidia-geforce-gtx-980-review/3
9. https://www.datascience.com/blog/cpu-gpu-machine-learning
10. R. Ronen, A. Mendelson, K. Lai, S.-L. Lu, F. Pollack, J. Shen, Coming challenges in microarchitecture and architecture. Proc. IEEE **89**(3), 325–340 (2001)
11. M. Flynn, P. Hung, K. Rudd, Deep submicron microprocessor design issues. IEEE Micro **19**(4), 11–22 (1999)
12. O.H.M. Ross, R. Sepulveda, *High Performance Programming for Soft Computing* (CRC Press, Boca Raton, 2014)
13. M. Gautschi, P.D. Schiavone, A. Traber, I. Loi, A. Pullini, D. Rossi, E. Flamand, F.K. Gurkaynak, L. Benini, Near-threshold RISC-V core with DSP extensions for scalable IoT endpoint devices. IEEE Trans. Very Large Scale Integr. Sys. **25**(10), 2700–2713 (2017)
14. A. Waterman, K. Asanovic, *The RISC-V Instruction Set Manual; Volume I: User-Level ISA* (SiFive Inc. and CS Division, EECS Department, University of California, Berkeley, 2017)
15. Sparc.org, *Technical Documents (SPARC Architecture Manual Version 8.)* (SPARC International, Inc., California). http://sparc.org/technical-documents/#V8. Accessed 22 Sep 2019
16. Von Neumann Architecture, teach-ict, https://www.teach-ict.com/2016/AS_Computing/OCR_H046/1_1_characteristics_components/111_architecture/von_neumann/miniweb/index.php. Accessed 7 Oct 2019
17. J.L. Hennessy, D.A. Patterson, *Computer Architecture: A Quantitative Approach*, 6th edn. (Elsevier, Amsterdam, 2017)
18. Modified Harvard Architecture, Wikipedia, https://en.wikipedia.org/wiki/Modified_Harvard_architecture. Accessed 9 Oct 2019

Chapter 2
Numerical Computing

2.1 Introduction

Any physical system can be represented mathematically by linear or nonlinear partial differential equations (PDEs). PDE involves two or more independent variables and it is used to represent 2D or 3D problems [1]. When a function involves one independent variable, the equation is called an ordinary differential equation (ODE) and it is used to represent 1D problems. The discretization of partial differential equations leads to large sparse or dense system of linear equations (SLEs) or system of nonlinear equations (SNLEs) as depicted in Fig. 2.1. Sometimes the number of theses equations can reach to 100 million or more [1].

In general, methods to solve PDEs and ODEs can be classified into two categories: analytical methods and iterative methods. Due to the increasing complexities encountered in the development of modern VLSI technology, analytical solutions usually are not feasible. So, iterative methods are used.

SNLEs are generally represented in $f(x) = 0$ form. A SNLEs solver must determine the values of x. Nonlinear equations can be in the form of cos, sin, log, polynomial function, parabolic function, exponential, and complex. They cover different domains such as materials, fluid, chemistry, and physics. In general, methods to solve SNLEs are iterative methods [1].

The most common iterative methods are Newton, Quai-Newton, Secant, and Muller. These methods are summarized in Fig. 2.1 [2].

SLEs are generally represented in a matrix form. Given a system of linear equations in the form of $Ax = b$, where x is the unknowns, there can be one solution, an infinite number of solutions, or no solution. A SLEs solver must determine the values of x for which the product with matrix (A) generates the vector constant (b). In general, methods to solve SLEs can be classified into two categories: direct methods and iterative methods. Both iterative and direct solvers are widely used [3–5].

Direct methods determine the exact solution in a single process while iterative methods start with an initial guess and compute a sequence of intermediate results

© Springer Nature Switzerland AG 2020
K. S. Mohamed, *Neuromorphic Computing and Beyond*,
https://doi.org/10.1007/978-3-030-37224-8_2

until it converges to a final solution. However, convergence is not always guaranteed for iterative methods since they are more sensitive to the matrix type being solved (Sparse, Dense, Symmetric, Square) which limits the scope of the range of the matrices that can be solved.

But, iterative methods are useful for finding solutions to large systems of linear equations where direct methods are prohibitively expensive in terms of computation time.

Direct methods are typically used for dense matrices of small to moderate size, which consist mainly of nonzero coefficients, as they would require large number of iterations and thus more memory accesses, if iterative methods are used. On the other hand, iterative methods are preferred for sparse matrices or dense matrices of large size, which are matrices that have a lot of zero coefficients, to make use of this special structure in reducing the number of iterations.

The most common direct methods are Gaussian elimination, Gauss–Jordan elimination, and LU decomposition. All of them have a computational complexity of order n^3 and are sensitive to numerical errors. The most common iterative methods are Jacobi, Gauss–Seidel, and Conjugate Gradient methods. They have computational complexity of order n^2 for each iteration step. These methods are summarized in Fig. 2.1.

Recently, advanced methods based on hybrid solutions and machine learning is proposed to solve SNLEs/SLEs. Hybrid methods combine new techniques with the conventional iterative methods.

ML-based methods such as genetic algorithm (GA) and artificial neural network (ANN) enhance solving process by accelerating the convergence of iterative numerical methods which reduces the computational efforts [6].

2.2   Numerical Analysis for Electronics

Since the advent of integrated circuits (ICs), numerical analysis has played an important role in simulating analog, mixed-signal, and RF designs. Circuit simulations, electromagnetic (EM) simulations, and device simulations are built using many numerical analysis methods [7–14]. The universe is continuous, but computers are discrete. Computers does not understand Phy/Maths equations, so we need to discretize/sample them to be able to program them.

2.2.1   Why EDA

As complexity of current day electronic design increases, manual design becomes unrealistic. Moreover, automation ensures fast time to market and fewer errors. VLSI cycle is shown in Fig. 2.2. Numerical methods are used at circuit design level. Logic design uses Booleans.

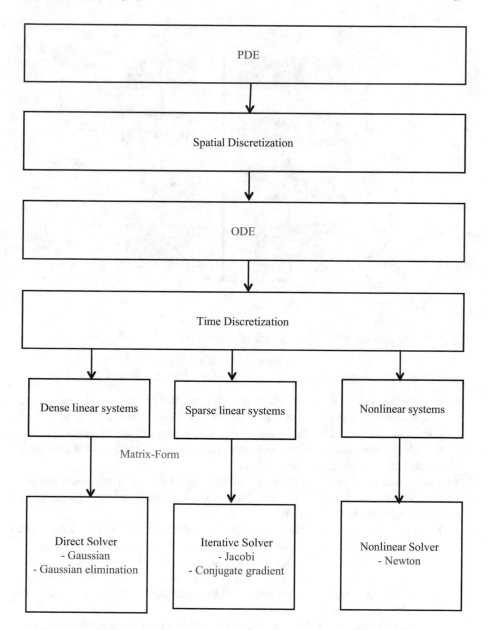

Fig. 2.1 Different methods of solving systems of linear equations: the big picture

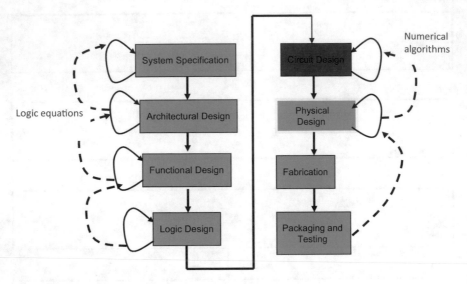

Fig. 2.2 VLSI cycle

2.2.2 Applications of Numerical Analysis

The field of numerical analysis includes many applications such as (Fig. 2.3):

- **Computing values of functions**: One of the simplest problems is the evaluation of a function at a given point. The most straightforward approach of just plugging in the number in the formula is sometimes not very efficient. For polynomials, a better approach is using the Horner scheme, since it reduces the necessary number of multiplications and additions.
- **Interpolation** solves the following problem: given the value of some unknown function at a number of points, what value does that function have at some other point between the given points?
- **Extrapolation** is very similar to interpolation, except that now the value of the unknown function at a point which is outside the given points must be found.
- **Regression** is also similar, but it takes into account that the data is imprecise. Given some points, and a measurement of the value of some function at these points (with an error), the unknown function can be found. The least squares method is one way to achieve this.
- **Solving equations and systems of equations**: Another fundamental problem is computing the solution of some given equation. Two cases are commonly distinguished, depending on whether the equation is linear or not.
- **Solving eigenvalues or singular value problems**: Several important problems can be phrased in terms of eigenvalue decompositions or singular value decompositions. For instance, the spectral image compression algorithm is based on the singular value decomposition. The corresponding tool in statistics is called principal component analysis.

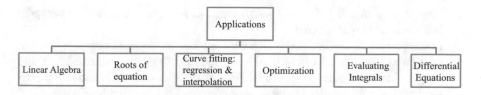

Fig. 2.3 Applications of numerical analysis

- **Optimization**: Optimization problems ask for the point at which a given function is maximized (or minimized). Often, the point also has to satisfy some constraints. The field of optimization is further split in several subfields, depending on the form of the objective function and the constraint. For instance, linear programming deals with the case that both the objective function and the constraints are linear. A famous method in linear programming is the simplex method.
- **Evaluating integrals**: Numerical integration, in some instances also known as numerical quadrature, asks for the value of a definite integral. Popular methods use one of the Newton–Cotes formulas (like the midpoint rule or Simpson's rule) or Gaussian quadrature. These methods rely on a "divide and conquer" strategy, whereby an integral on a relatively large set is broken down into integrals on smaller sets. In higher dimensions, where these methods become prohibitively expensive in terms of computational effort, one may use Monte Carlo or quasi-Monte Carlo methods (see Monte Carlo integration), or, in modestly large dimensions, the method of sparse grids.
- **Differential equations**: Numerical analysis is also concerned with computing (in an approximate way) the solution of differential equations, both ordinary differential equations and partial differential equations.

2.2.3   Approximation Theory

Use computable functions $p(x)$ to approximate the values of functions $f(x)$ that are not easily computable or use approximations to simplify dealing with such functions. The most popular types of computable functions $p(x)$ are polynomials, rational functions, and piecewise versions of them, for example, spline functions. Trigonometric polynomials are also a very useful choice.

- Best approximations. Here a given function $f(x)$ is approximated within a given finite-dimensional family of computable functions. The quality of the approximation is expressed by a functional, usually the maximum absolute value of the approximation error or an integral involving the error. Least squares approximations and minimax approximations are the most popular choices.
- Interpolation. A computable function $p(x)$ is to be chosen to agree with a given $f(x)$ at a given finite set of points x. The study of determining and analyzing such

interpolation functions is still an active area of research, particularly when $p(x)$ is a multivariate polynomial.

- Fourier series. A function $f(x)$ is decomposed into orthogonal components based on a given orthogonal basis $\{\varphi 1, \varphi 2, \ldots\}$, and then $f(x)$ is approximated by using only the largest of such components. The convergence of Fourier series is a classical area of mathematics, and it is very important in many fields of application. The development of the Fast Fourier Transform in 1965 spawned a rapid progress in digital technology. In the 1990s wavelets became an important tool in this area.

- Numerical integration and differentiation. Most integrals cannot be evaluated directly in terms of elementary functions, and instead they must be approximated numerically. Most functions can be differentiated analytically, but there is still a need for numerical differentiation, both to approximate the derivative of numerical data and to obtain approximations for discretizing differential equations.

2.3   Different Methods for Solving PDEs and ODEs

Partial and ordinary differential equations are used in modeling and solving many problems in various fields such as physics, chemistry, biological, and mechanics (Table 2.1).

Differential equations represent the rate of change in these systems and they are classified according to their **order**. For example, a first-order equation includes a first derivative as its highest derivative. Also, it can be **linear** or nonlinear. Some of physical problems are governed by first-order PDEs while numerous problems are governed by second-order PDEs. Only few problems are governed by higher order PDEs. Higher order equations can be reduced to a system of first-order equations, by redefining a variable.

Moreover, a linear differential equation is **homogeneous** if every term contains the dependent variable or its derivatives. An inhomogeneous differential equation has at least one term that contains no dependent variable.

The general form for the second-order differential equation is given by:

$$A\frac{\partial^2 u}{\partial x^2} + B\frac{\partial^2 u}{\partial x \partial y} + C\frac{\partial^2 u}{\partial y^2} + D\frac{\partial u}{\partial x} + E\frac{\partial u}{\partial y} + Fu = G \qquad (2.1)$$

where A, B, C, D, E, F, and G are either real constants or real-valued functions of x and/or y. The different types of PDEs and the different solutions for them are summarized in Table 2.2.

There are many methods to solve PDEs and ODEs. We need to specify the solution at $t = 0$, or $x = 0$, i.e., we need to specify the **initial conditions** and the **boundary values**. Analytical methods are based on proposing a trial solution or using separation of variable then integration or choosing a basis set of functions with adjustable parameters and proceed approximating the solution by varying these parameters.

Table 2.1 Examples for differential equations

Domain	Example	Differential equation
Electromagnetic	Wave equation	Linear $$\nabla^2 u = \frac{1}{c^2}\frac{\partial^2 u}{\partial t^2}$$ Nonlinear $$\nabla^2 u = \frac{1}{c^2}\frac{\partial^2 u}{\partial t^2}$$
	Poisson's equation	$$\nabla^2 u = -\frac{\rho}{\varepsilon_0}$$
	Laplace's equation	$\nabla^2 u = 0$
	Maxwell equations	$$\nabla \times \vec{E} = -\mu\frac{\partial \vec{H}}{\partial t}$$ $$\nabla \times \vec{H} = -\varepsilon\frac{\partial \vec{E}}{\partial t} + \sigma \times \vec{E}$$ $$\nabla \cdot \varepsilon\vec{E} = \rho$$ $$\nabla \cdot \mu\vec{H} = 0$$
Thermodynamics	Diffusion equation	$$\nabla^2 u = \frac{1}{h^2}\frac{\partial u}{\partial t}$$
Quantum mechanics	Schrödinger's equation	$$-\frac{\hbar^2}{2m}\nabla^2 u + Vu = i\hbar\frac{\partial u}{\partial t}$$
Mechanics	Newton's Laws	$$F = ma = m\frac{d^2 x}{dt^2}$$

Table 2.2 The different types of PDEs and the different solutions

	PDE type	Solution type	Example
$B^2 - 4AC < 0$	Elliptic	Equilibrium or steady state	Poisson equation
$B^2 - 4AC = 0$	Parabolic	Solution "propagates" or diffuses	Heat equation
$B^2 - 4AC > 0$	Hyperbolic	Solution propagates as a wave	Wave equation

But, due to the increasing complexities encountered in the development of modern VLSI technology, analytical solutions usually are not feasible. So, iterative methods are used. The explanation of the methods will be applied on boundary value problems partial differential equation with Dirichlet conditions.

2.3.1 Iterative Methods for Solving PDEs and ODEs

2.3.1.1 Finite Difference Method (Discretization)

FDM is a numerical method for solving differential equations by approximating them with difference equations, in which finite differences approximate the derivatives. FDMs are thus discretization methods. FDMs convert a linear (nonlinear) ODE (Ordinary Differential Equations)/PDE (Partial differential equations) into a system of linear (nonlinear) equations, which can then be solved by matrix algebra techniques. The reduction of the differential equation to a system of algebraic equations makes the problem of finding the solution to a given ODE ideally suited to modern computers, hence the widespread use of FDMs in modern numerical analysis. The finite difference method is also one of a family of methods for approximating the solution of partial differential equations such as heat transfer, stress/strain mechanics problems, fluid dynamics problems, and electromagnetics problems. [15].

By discretizing the domain into grid of spaced points, the first derivative and the second derivative are approximated using the following formulas:

$$\frac{du}{dx} = \frac{u_{n+1} - u_n}{\Delta x} \text{ Forward Euler Method} \tag{2.2}$$

$$\frac{du}{dx} = \frac{u_n - u_{n-1}}{\Delta x} \text{ Backward Euler Method} \tag{2.3}$$

$$\frac{d^2 u}{dx^2} = \frac{u_{n-1} - 2u_n + u_{n+1}}{\Delta x^2} \tag{2.4}$$

where Δx is the **step size**, after that solve the resultants system of linear or nonlinear equations.

2.3.1.2 Finite Element Method (Discretization)

Arbitrary-shaped boundaries are difficult to implement in finite difference methods. So, finite element method (FEM) is used. FEM covers the space with finite elements. The elements do not need to have the same size and shape. We get function value and derivative by interpolation. FEM is a numerical method that is used to solve boundary value problems defined by a partial differential equation (PDE) and a set of boundary conditions. The first step of using FEM to solve a PDE is discretizing the computational domain into finite elements. Then, the PDE is rewritten in a weak formulation. After that, proper finite element spaces are chosen and the finite element scheme is formed from the weak formulation. The next step is calculating those element matrices on each element and assembling the element matrices to form a global linear system. Then, the boundary conditions are applied, the sparse linear system is solved, and finally post-processing of the numerical solution is done [16, 17].

2.3.1.3   Legendre Polynomials

Legendre Polynomials are solutions to Legendre's ODE which can be expressed as follows [18, 19]:

$$\left(1-x^2\right)\frac{dy^2}{d^2x} - 2x\frac{dy}{dx} + k\left(k+1\right)y = 0 \tag{2.5}$$

The notation for expressing Legendre Polynomials is P_k and examples of the first 7 Legendre Polynomials are:

$$P_0\left(x\right) = 1 \tag{2.6}$$

$$P_1\left(x\right) = x \tag{2.7}$$

$$P_2\left(x\right) = \frac{1}{2}\left(3x^2 - 1\right) \tag{2.8}$$

$$P_3\left(x\right) = \frac{1}{2}\left(5x^3 - 3x\right) \tag{2.9}$$

$$P_4\left(x\right) = \frac{1}{8}\left(35x^4 - 30x^2 + 3\right) \tag{2.10}$$

$$P_5\left(x\right) = \frac{1}{8}\left(63x^5 - 70x^3 + 15x\right) \tag{2.11}$$

$$P_6\left(x\right) = \frac{1}{16}\left(231x^6 - 315x^4 + 105x^2 - 5\right) \tag{2.12}$$

The solution to Legendre's ODE is a series solution and can be written as

$$y = \sum_{n=0}^{\infty} a_n x^n \tag{2.13}$$

The real question is how we can get the values of a_n. By following a long-term derivation we can reach a recurrence relation that shows that a value of a_n depends on the values 2 steps before it, which means that if we have a_0 and a_1 we can derive the rest of the series.

$$a_{n+2} = \frac{a_n(n-k)(n+k+1)}{(n+2)(n+1)} \tag{2.14}$$

When we start expanding these coefficients, we notice that the resultant series can be written as 2 series, one with odd exponents and the other with even exponents. If k is odd, then the odd series converges and the even series diverges. On the other hand, if k is even, then the odd series diverges and the even series converges. The convergent series can then be used as the as the Legendre Polynomial P_k for this k. A general closed form for computing Legendre Polynomials has been derived and is as follows:

$$P_k(x) = \sum_{n=0}^{\text{LOOP\_LIMIT}} (-1)^n \frac{(2k-2n)!}{2^k n!(k-n)!(k-2n)!} x^{k-2n} \tag{2.15}$$

The parameter LOOP _ LIMIT can be computed as follows:

$$\text{LOOP\_LIMIT} = \begin{cases} k/2, & \text{if } k \text{ is even} \\ (k-1)/2, & \text{if } k \text{ is odd} \end{cases} \tag{2.16}$$

2.3.2   Hybrid Methods for Solving PDEs and ODEs

In [20], the authors propose a novel hybrid analytical numerical method for solving certain linear PDEs. The new method has advantages in comparison with classical methods, such as avoiding the solution of ordinary differential equations that result from the classical transforms, as well as constructing integral solutions in the complex plane which converge exponentially fast and which are uniformly convergent at the boundaries.

2.3.3   ML-Based Methods for Solving ODEs and PDEs

The authors in [21] present a novel approach based on modified artificial neural network and optimization technique to solve partial differential equations. In [22], the authors propose using genetic algorithm to solve PDEs. Genetic algorithm is based on iterative procedures of search for an optimal solution for a problem which have multiple local minima or maxima. The algorithm passes through steps of recombination including crossover, and mutation then selection which increase the probability of finding the most optimum solution, which is the reduced model with the least error compared to the original transfer model. The error is compared with the original transfer function in terms of fitness function. Before applying the genetic operators, a method of encoding should be chosen to represent the data either in float form which is the raw form of the data or binary representation or any other representation. The crossover operator is a process of creating new individuals by selecting two or more parents and passing them through crossover procedures producing one or two or more individuals. Unlike real life there is no obligation to be abided by nature rules, so the new individual can have more than two parents. There is more than one method for crossover process like simple crossover which includes exchange of genes between the two chromosomes according to a specified crossover rate. Arithmetic crossover occurs by choosing two or more individuals randomly from the current generation and multiplying them by one random in the case of the inevitability of the presence of a certain defined domain and more than one random if there is no physical need for a defined search domain. The second process of genetic operators is mutation which is a process that occurs to prevent falling of all solutions of the population into a local optimum of the problem. The third genetic operator is selection which is the process of choosing a certain number of individuals for the next generation to be recombined generating new individuals aiming to find the most optimum solution. There is more than one technique to select individuals for the next generation. The Elitism selection is simply selecting the fittest individuals from the current population for the next population. This method guarantees a high probability of getting closer to the most optimum solution due to passing of the fittest chromosomes to the crossover operator producing fitter individuals. Another technique is Roulette wheel selection; in this kind of selection the parenthood probability is directly proportional to the fitness of the individuals where every individual is given a weight proportional to its fitness having a higher probability to be chosen as a parent for the next-generation individuals and this technique is similar to rank selection where all individuals in the current population are ranked according to their fitness. Each individual is assigned a weight inversely proportional to the rank. The fitness function is the common evaluation factor among all selection techniques. Fitness function is a function which evaluates the fitness of each individual to be selected for the next generation [23].

2.3.4   How to Choose a Method for Solving PDEs and ODEs

As stated earlier, there exists a variety of algorithms for solving PDEs and ODEs (Fig. 2.4). Selecting the best solver algorithm depends on number of iterations and memory usage. Comparison between different methods in terms of computation time and number of iterations is shown in Table 2.3. The criteria on which we can choose which method to be used is summarized in Table 2.4.

Fig. 2.4 Different
methods of solving PDEs
and ODEs

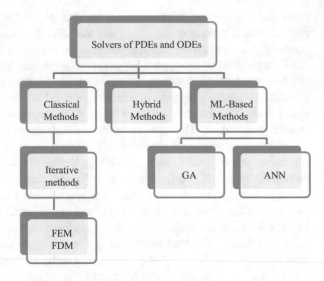

Table 2.3 Comparison between different iterative methods

Methods	Computation time (Sec)	Number of iterations	Memory usage
FEM	7	100	High
FDM	9	170	High

Table 2.4 How to choose a
method for solving nonlinear
equations

PDEs and ODEs solver methods	Boundaries	
	Arbitrary shaped	Rectangular
FEM	✓	✓
FDM	✗	✓

2.4   Different Methods for Solving SNLEs

The nonlinear systems problems are important in physics and engineering domains. For example, device simulation involves solving a large system of nonlinear equations. So, most simulation time is spent for solving this huge number of equations.

A nonlinear system of equations is a system in which at least one of the equations is nonlinear. For example, a system that contains one quadratic equation and one linear equation is a nonlinear system. A system made up of a linear equation and a quadratic equation can have no solution, one solution, or two solutions, as shown below in Fig. 2.5. Nonlinear systems can be solved **analytically by graphing, substitution, or eliminations.**

The substitution method is a good choice when equation is solved for a variable, both equations are solved for the same variable, or a variable in either equation has a coefficient of 1 or −1.

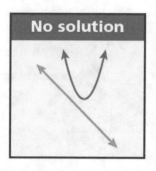  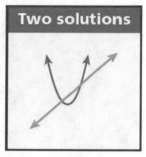

Fig. 2.5 Solution of linear and nonlinear equations

The elimination method is a good choice when both equations have the same variable term with the same or opposite coefficients or when a variable term in one equation is a multiple of the corresponding variable term in the other equation. In this work, we focus on the numerical methods, not the analytical methods.

A lot of techniques that are used for nonlinear systems come from linear systems, because nonlinear systems can sometime be approximated by linear systems.

A nonlinear equation involves terms of degree higher than one. Any nonlinear equation can be expressed as

$$f(x) = 0$$

They cannot be solved using direct methods, but using iterative methods. In this section, all methods to solve SNLEs are discussed. The explanation of the methods will be on 3 variables nonlinear equations shown below:

$$3x_1 - \cos x_2 x_3 - 0.5 = 0$$
$$x_1^2 - 81(x_2 + 0{,}1)^2 + \sin x_3 + 1.06 = 0 \tag{2.17}$$
$$e^{-x_2 x_1} + 20x_3 + \frac{10\Pi - 3}{3} = 0$$

2.4.1 Iterative Methods for Solving SNLEs

2.4.1.1 Newton Method and Newton–Raphson Method

Nonlinear problems are often treated numerically by reducing them to a sequence of linear problems. As a simple but important example, consider the problem of solving a nonlinear equation $f(x) = 0$. Approximate the graph of $y = f(x)$ by the tangent line at a point $x(0)$ near the desired root and use the root of the tangent line to approximate the root of the original nonlinear function $f(x)$. This leads to Newton's

Fig. 2.6 Newton method

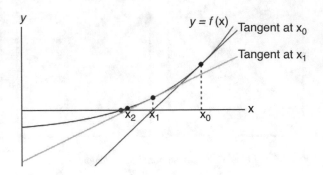

iterative method for finding successively better approximations to the desired root as shown in Fig. 2.6.

This method originates from the **Taylor's series** expansion of the function $f(x)$ about the point x_1 as follows [24]:

$$f(x) = f(x_1) + (x - x_1)f'(x_1) + \frac{1}{2!}(x - x_1)^2 f''(x_1) + \ldots \approx f(x_1) + (x - x_1)f'(x_1) = 0 \quad (2.18)$$

Rearrange Eq. (2.18) results in Eq. (2.19):

$$x = x_1 - \frac{f(x_1)}{f'(x_1)} \quad (2.19)$$

Generalizing Eq. (2.19) we obtain Newton's iterative method that can only be used to solve nonlinear equations involving only a single variable:

$$x_{n+1} = x_n - \frac{f(x_n)}{f'(x_n)} \quad (2.20)$$

To be able to use Eq. (2.20) in solving nonlinear equations involving many variables, we define the Jacobian matrix is a matrix of first-order partial derivatives. This is called Newton–Raphson method. So, Newton method is used to solve single variable nonlinear equations, while Newton–Raphson method is used to solve multivariable nonlinear equations [25].

$$J(x) = \begin{bmatrix} \dfrac{df_1}{dx_1} & \dfrac{df_1}{dx_2} & \dfrac{df_1}{dx_3} \\[2mm] \dfrac{df_2}{dx_1} & \dfrac{df_2}{dx_2} & \dfrac{df_2}{dx_2} \\[2mm] \dfrac{df_3}{dx_1} & \dfrac{df_3}{dx_2} & \dfrac{df_3}{dx_2} \end{bmatrix} \qquad (2.21)$$

So, the following iterative equation is used for solving nonlinear equations involving many variables.

$$x_{n+1} = x_n - J(x_0)^{-1} f(x_n) \qquad (2.22)$$

The detailed steps of the solution of Eq. (2.17) can be shown as follows:

- Step 1: Let $x_1 = x_2 = -x_3 = 0.1$.
- Step 2: Find $F(x)$

$$F(x) = \begin{bmatrix} 3x_1 - \cos x_2 x_3 - 0.5 \\[2mm] x_1^2 - 81(x_2 + 0{,}1)^2 + \sin x_3 + 1.06 \\[2mm] e^{-x_2 x_1} + 20x_3 + \dfrac{10\Pi - 3}{3} \end{bmatrix} \qquad (2.23)$$

- Step 3: Find the Jacobi Matrix

$$J(x) = \begin{bmatrix} 3 & x_3 \sin x_2 x_3 & x_2 \sin x_2 x_3 \\[2mm] 2x_1 & -162(x_2 + 0.1) & \cos x_3 \\[2mm] -x_2 e^{-x_2 x_1} & -x_1 e^{-x_2 x_1} & 20 \end{bmatrix} \qquad (2.24)$$

- Step 4: Find $F(x_0)$ and $J(x_0)$

$$F(x_0) = \begin{bmatrix} 0.3 - \cos - 0.01 - 0.5 \\[2mm] 0.01 - 3.24 + \sin - 0.1 + 1.06 \\[2mm] e^{-0.01} - 2 + \dfrac{10\Pi - 3}{3} \end{bmatrix} \qquad (2.25)$$

$$J(x_0) = \begin{bmatrix} 3 & -0.1\sin - 0.01 & 0.1\sin - 0.01 \\[2mm] 0.2 & -32.4) & \cos - 0.1 \\[2mm] -0.1e^{-0.01} & -0.1e^{-0.01} & 20 \end{bmatrix} \qquad (2.26)$$

- Step 5: Apply into the following equation:

$$x_1 = x_0 - J(x_0)^{-1} f(x_0) \qquad (2.27)$$

- The solution of this step is $\begin{bmatrix} 0.5 \\ 0.01 \\ -0.5 \end{bmatrix}$

- Step 6: Use the results of x_1 to find our next iteration x_2 by using the same procedure.
- The final solution is $\begin{bmatrix} 0.5 \\ 0 \\ -0.5 \end{bmatrix}$

One of the advantages of Newton's method is its simplicity. The major disadvantage associated with Newton's method is that $J(x)$, as well as its inversion, has to be calculated for each iteration which consumes time depending on the size of your system. The convergences of the classical solvers such as Newton-type methods are highly sensitive to the initial guess of the solution. So, it might not converge if the initial guess is poor. Newton method flowchart is shown in Fig. 2.7. Newton–Raphson is used to find the reciprocal of D and multiply that reciprocal by N to find the final quotient Q.

2.4.1.2 Quasi-Newton Method *aka* Broyden's Method

It uses an approximation matrix that is updated at each iteration instead of the Jacobian matrix. So, the iterative procedure for Broyden's method is the same as Newton's method except that an approximation Matrix A is used instead of Jacobi method. The main advantage of Broyden's method is the reduction of computations, but it needs more iterations than Newton's method [26, 27]. Quasi-Newton method is a method used to either find zeroes or local maxima and minima of functions, as an alternative to Newton's method. As the most important disadvantage of Newton's method is the requirement that $F'(xk)$ be determined for each k (number of iteration), this is a very costly operation, but the exact cost varies from problem to problem based on its complexity. It works as follows:

- Step 1: Given a starting point $\mathbf{x}(0) \in \mathbf{Rn}$, $\mathbf{H(0)} > 0$. Where $\mathbf{H}$ is the Hessian matrix of the equation, $\mathbf{H}^{-1}$ is the inverse of the Hessian matrix, and $\mathbf{t}$ is the step size. $\mathbf{H}$ is given by Eq. (2.28)
- Step 2: For k (iteration number until reaching an acceptable error of x) = 1, 2, 3, …, repeat:

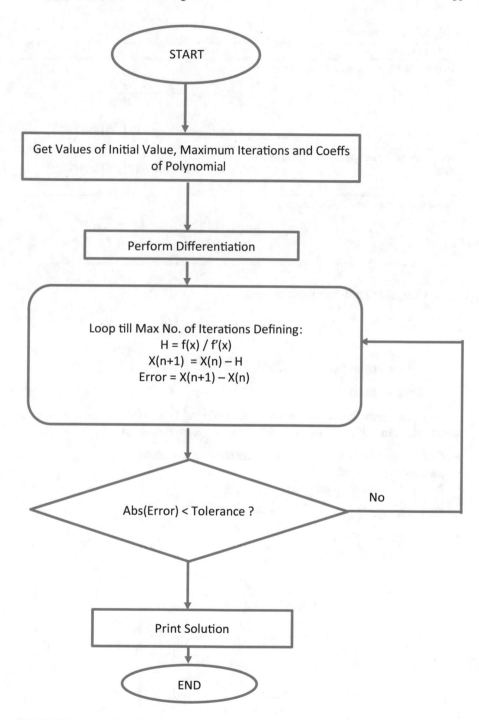

Fig. 2.7 Newton method flowchart

Table 2.5 H update methods

Update Method	Equation
Broyden-Fletcher-Goldfarb-Shanno (BFGS) update	$H_k = H_{k-1} + \dfrac{yy^T}{y^T s} - \dfrac{H_{k-1}ss^T H_{k-1}}{s^T H_{k-1}s}$ $s = x^{(k)} - x^{(k-1)}, \quad y = \nabla f\left(x^{(k)}\right) - \nabla f\left(x^{(k-1)}\right)$
Symmetric rank one update: (B is the Hessian matrix)	$s = x^{(k)} - x^{(k-1)}, \quad y = \nabla f\left(x^{(k)}\right) - \nabla f\left(x^{(k-1)}\right)$
Davidon-Fletcher-Powell (DFP) update:	$H_k = \left(I - \dfrac{ys^T}{s^T y}\right)H_{k-1}\left(I - \dfrac{sy^T}{s^T y}\right) + \dfrac{yy^T}{s^T y}$

1. Compute Quasi-Newton direction:

$$\Delta x(k-1) = -\nabla f(x(k-1) * H^{-1}(k-1)$$

2. Update $x(k) = x(k-1) + t\Delta x(k-1)$.

3. Compute $\mathbf{H}(k)$.

Different methods use different rules for updating $\mathbf{H}$ in step 3. These methods are summarized in Table 2.5. Advantages of Quasi-Newton method:

- Computationally cheap compared with Newton's method.
- Faster computation.
- No need for second derivative.

Disadvantages:

- More convergence steps.
- Less precise convergence path.

$$\mathbf{H} = \begin{bmatrix} \dfrac{\partial^2 f}{\partial x_1^2} & \dfrac{\partial^2 f}{\partial x_1 \partial x_2} & \cdots & \dfrac{\partial^2 f}{\partial x_1 \partial x_n} \\[2ex] \dfrac{\partial^2 f}{\partial x_2 \partial x_1} & \dfrac{\partial^2 f}{\partial x_2^2} & \cdots & \dfrac{\partial^2 f}{\partial x_2 \partial x_n} \\[1ex] \vdots & \vdots & \ddots & \vdots \\[1ex] \dfrac{\partial^2 f}{\partial x_n \partial x_1} & \dfrac{\partial^2 f}{\partial x_n \partial x_2} & \cdots & \dfrac{\partial^2 f}{\partial x_n^2} \end{bmatrix}. \tag{2.28}$$

Fig. 2.8 Secant line for the first two guesses

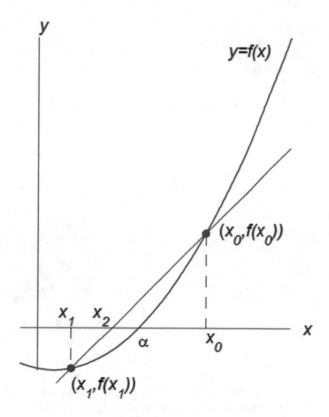

2.4.1.3   The Secant Method

In this method, we replace the derivative $f'(x_i)$ in the Newton–Raphson method with the following linearization equation [28, 29]:

$$f'(x_{n+1}) = \frac{f(x_n) - f(x_{n+1})}{x_n - x_{n+1}} \tag{2.29}$$

Moreover, in contrast to Newton–Raphson method, here we need to provide two initial values of x to get the algorithm started (x_0, x_1). In other words, we approximate the function by a straight line (Figs. 2.8 and 2.9).

The absolute error in every iteration can be calculated as

$$e = |x_n - x_{n-1}| \tag{2.30}$$

Iteration is stopped if one of the following conditions is met:

- Number of iterations exceeds a certain number.
- Error is below a certain threshold.

Fig. 2.9 Secant line for
the new estimate with the
second guess from
previous iteration

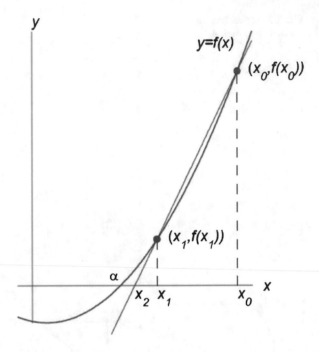

However, the secant method is not always guaranteed to converge. There is no guarantee that the secant method will converge to the root in the following scenarios:

- If the initial estimates are not close enough to the root.
- If the value of the first derivative of the function at the root equals zero.

Advantages:

- Does not require the calculation of a derivative, unlike Newton–Raphson.
- Relatively fast rate of convergence (faster than linear, but slower than the Newton–Raphson method).

Disadvantages:

- Requires two initial guesses not one like the Newton–Raphson method.
- May not converge (if the conditions mentioned above are not met).

2.4.1.4 The Muller Method

It is based on Secant method, but instead of linear approximation, it uses quadratic approximate. It works only for a set of polynomial equations. It begins with three initial assumptions of the root, and then constructing a parabola through these three points, and takes the intersection of the x-axis with the parabola to be the next approximation. This process continues until a root with the desired level of accuracy is found [30].

2.4.2   Hybrid Methods for Solving SNLEs

The iterative methods are categorized as either locally convergent or globally convergent. The locally convergent methods have fast convergence rate, but they require initial approximations close to the roots. While the globally convergent methods have very slow convergence rate and initial approximations far from the roots should lead to convergence.

Hybrid methods combine locally and globally convergent methods in the same algorithm such that if the initial approximations are far from the roots, the slow globally convergent algorithm is used until the points are close enough to the roots, then the locally convergent method is used.

In [31], the authors propose a hybrid method for solving systems of nonlinear equations that require fewer computations than those using regular methods. The proposed method presents a new technique to approximate the Jacobian matrix needed in the process, which still retains the good features of Newton's type methods, but which can reduce the run time.

In [32], authors propose a hybrid approach for solving systems of nonlinear equations. The approach is based on using chaos optimization technique with quasi-Newton method to speed-up run time.

2.4.3   ML-Based Methods for Solving SNLEs

The authors in [33] present a novel framework for the numerical solution of nonlinear differential equations using neural networks with back-propagation algorithm.

In [6], the authors propose using genetic algorithm to solve SNLEs. GA approach optimizes the space and time complexities in solving the nonlinear system as compared to the traditional computing methods.

2.4.4   How to Choose a Method for Solving Nonlinear Equations

As stated earlier, there exists a variety of algorithms for solving nonlinear systems (Fig. 2.10). Selecting the best solver algorithm depends on number of iterations and memory usage.

Comparison between different methods based on Eq. (2.31) in terms of computation time and number of iterations is shown in Table 2.6.

One of the advantages of Newton's method is its simplicity. The major disadvantage associated with Newton's method is that $J(x)$, as well as its inversion, has to be calculated for each iteration which consume time depending on the size of your system. The main advantage of Broyden's method is the reduction of computations, but it needs more iterations than Newton's method. The criteria on which we can choose which method to be used is summarized in Table 2.7.

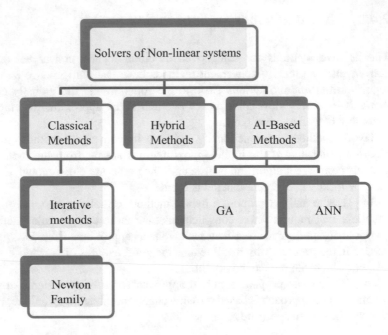

Fig. 2.10 Different methods of solving systems of nonlinear equations

Table 2.6 Comparison between different iterative methods

Methods	Computation time (s)	Number of iterations	Memory usage
Newton	3	13	The highest
Quasi-Newton	5	20	Less
Secant	4	10	Less
Muller	4	9	Less

Table 2.7 How to choose a method for solving nonlinear equations

SNLEs solver methods	Jacobian matrix type		Converge
	Sparse	Dense	
Newton	✓	✗	Locally
Quasi-Newton	✗	✓	Locally
Secant	✗	✓	Globally
Muller	✗	✓	Globally

2.5   Different Methods for Solving SLEs

The linear systems problems are important in physics and engineering domains. For example, circuit simulation involves solving a large system of nonlinear equations. So, most simulation time is spent for solving this huge number of equations. A linear systems of equations can be represented as $Ax = b$.

It is important to consider that not all matrices can be solved using iterative methods. A matrix can be solved only when it is diagonally dominant. A matrix is said to be diagonally dominant if the magnitude of every diagonal entry is more than the sum of the magnitude of all the nonzero elements of the corresponding row. Both methods sometimes converge even if this condition is not satisfied. However, it is necessary that the magnitude of diagonal terms in a matrix is greater than the magnitude of other terms. A determinant equal zero means that the matrix is singular and the system is ill-conditioned [5]. In this section, all methods to solve SLEs are discussed. The explanation of the methods will be on a 3 x 3 linear equations shown below:

$$\begin{bmatrix} 4 & 2 & 3 \\ 3 & -5 & 2 \\ -2 & 3 & 8 \end{bmatrix} \begin{bmatrix} x \\ y \\ z \end{bmatrix} = \begin{bmatrix} 8 \\ -14 \\ 27 \end{bmatrix} \tag{2.31}$$

2.5.1 Direct Methods for Solving SLEs

2.5.1.1 Cramer's Rule Method

In linear algebra, Cramer's rule is an explicit formula for the solution of systems of linear equations with as many equations as unknowns; it is valid only when the system has a unique solution. It expresses the solution in terms of the determinants of the square coefficient matrix and of matrices obtained from it by replacing one column by the vector of right-hand sides of the equations [22].

The value of the determinant T should be checked because if $T = 0$, the equations wouldn't have a unique solution. The main advantage of this design is that it is very fast compared to other methods as it needs fewer number of clock cycles to calculate the result. The main disadvantage of this design is that it consumes a lot of resources. As a result, it is not a scalable solver. Cramer method is preferable to be used with square matrix. It relies on finding the inverse of matrix A. So, the solution to $Ax = b$ is $x = A^{-1}b$ [34–36].

The detailed steps of the solution of Eq. (2.31) can be shown as follows:

- Step 1: Write down the main matrix and find its determinant Δ.
- Step 2: Replace the first column of the main matrix with the solution vector (b) and find its determinant Δ_1.
- Step 3: Replace the second column of the main matrix with the solution vector and find its determinant Δ_2.
- Step 4: Replace the third column of the main matrix with the solution vector and find its determinant Δ_3.

- Step 5: Calculate the output using the following formulas $x = \dfrac{\Delta 1}{\Delta}$; $y = \dfrac{\Delta 2}{\Delta}$, $z = \dfrac{\Delta 3}{\Delta}$.

- Step 6: Solution = $\begin{bmatrix} -1 \\ 3 \\ 2 \end{bmatrix}$.

Cramer's rule is computationally inefficient for systems of more than two or three equations. In the case of n equations in n unknowns, it requires computation of $n + 1$ determinants, while Gaussian elimination produces the result with the same computational complexity as the computation of a single determinant. Cramer's rule can also be numerically unstable even for 2×2 systems. However, it has recently been shown that Cramer's rule can be implemented in $O(n^3)$ time, which is comparable to more common methods of solving systems of linear equations, such as Gaussian elimination (consistently requiring 2.5 times as many arithmetic operations for all matrix sizes), while exhibiting comparable numeric stability in most cases. In linear algebra Cramer's rule is an explicit formula for the solution of a system of linear equation. Cramer's rule flowchart is shown in Fig. 2.11.

The determinant is calculated in a recursive function that reduces the input matrix to multiple smaller matrices and calls itself with each of these matrices, until the size of these smaller matrices becomes 2×2, then it computes its determinant and returns it to the previous function call, which uses these values to obtain the result of the original input matrix. Determinant calculation example is shown in Fig. 2.12.

2.5.1.2 Gaussian Elimination Method

Gaussian elimination (also known as row reduction) is an algorithm for solving systems of linear equations. It is usually understood as a sequence of operations performed on the corresponding matrix of coefficients. This method can also be used to find the rank of a matrix, to calculate the determinant of a matrix, and to calculate the inverse of an invertible square matrix. To perform row reduction on a matrix, one uses a sequence of elementary row operations to modify the matrix until the lower left-hand corner of the matrix is filled with zeros, as much as possible. There are three types of elementary row operations [37, 38]:

- Swapping two rows
- Multiplying a row by a nonzero number
- Adding a multiple of one row to another row

Using these operations, a matrix can always be transformed into an upper triangular matrix, and in fact one that is in row echelon form. Once all of the leading coefficients (the leftmost nonzero entry in each row) are 1, and every column containing a leading coefficient has zeros elsewhere, the matrix is said to be in reduced row echelon form. This final form is unique; in other words, it is independent of the sequence of row operations used.

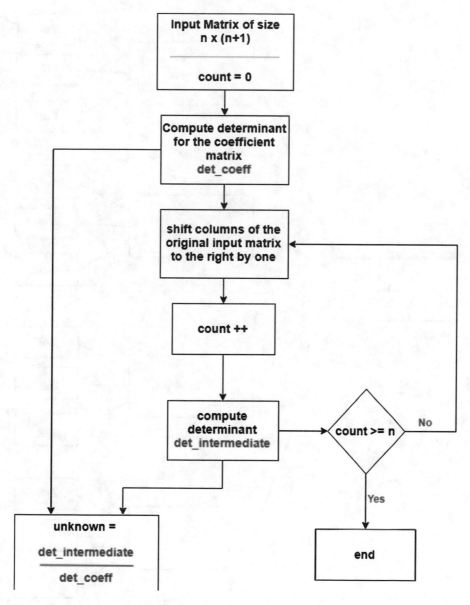

Fig. 2.11 Cramer's method flowchart

In gauss elimination, forward elimination is used first to eliminate all the elements in the lower triangle of the matrix. Next, backward elimination is used to eliminate all the elements in the upper triangle of the matrix, leaving only the main diagonal. The main disadvantage of this design is that it consumes a lot of logic resources, but it is fast [21, 22]. The detailed steps of the solution of Eq. (2.31) can be shown as follows [34, 35]:

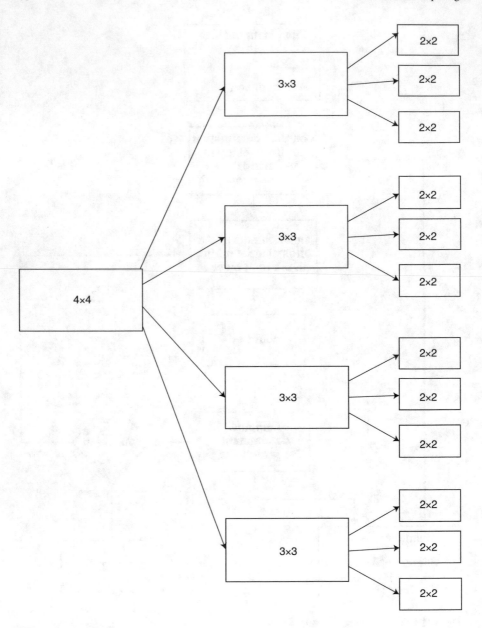

Fig. 2.12 Determinant calculation example

- Step 1: Make the pivot in the first column. In other words, obtain row echelon form for the matrix.
- Step 2: Make the pivot in the second column.
- Step 3: Make the pivot in the third column.

- Step 4: The output can be found from the matrix which has all ones as diagonal elements and zeros in lower triangles.

- Step 5: Solution = $\begin{bmatrix} -1 \\ 3 \\ 2 \end{bmatrix}$.

2.5.1.3 Gauss–Jordan (GJ) Elimination Method

GJ elimination is a method based on Gaussian elimination that puts zeros above and below each pivot giving in this way the inverse matrix. The differences from Gaussian elimination method are: When an unknown is eliminated from an equation, it is also eliminated from all other equation and all rows are normalized by dividing them by their pivot element. Hence, the elimination step results in an identity matrix rather than a triangular matrix. Therefore, back substitution is not necessary [36, 39]. The detailed steps of the solution of Eq. (2.31) can be shown as follows:

- Step 1: Make the pivot in the first column, then eliminate it. In other words, obtain reduced row echelon form for the matrix, where the lower left part of this matrix contains only zeros and all of the zero rows are below the nonzero rows. The matrix is reduced to this form by the elementary row operations: swap two rows, multiply a row by a constant, add to one row a scalar multiple of another.
- Step 2: Make the pivot in the second column, then eliminate it.
- Step 3: Make the pivot in the third column, then eliminate it.

- Step 4: The output can be found from the unity matrix.

- Step 5: Solution = $\begin{bmatrix} -1 \\ 3 \\ 2 \end{bmatrix}$.

Row reduction is the process of performing row operations to transform any matrix into (reduced) row echelon form. In reduced row echelon form, each successive row of the matrix has fewer dependencies than the previous, so solving systems of equations is a much easier task. The idea behind row reduction is to convert the matrix into an "equivalent" version in order to simplify certain matrix computations. Its two main purposes are to solve system of linear equations and calculate the inverse of a matrix. Carl Friedrich Gauss championed the use of row reduction, to the extent that it is commonly called **Gaussian elimination**. It was further popularized by Wilhelm Jordan, who attached his name to the process by which row reduction is used to compute matrix inverses, **Gauss–Jordan elimination**. A system of equations can be represented in a couple of different matrix forms. One way is to realize the system as the matrix multiplication of the coefficients in the system and the column vector of its variables. The square matrix is called the **coefficient matrix** because it consists of the coefficients of the variables in the system of equations. An alternate representation called an **augmented matrix** is created by stitching the columns of matrices together and divided by a vertical bar. The coefficient matrix is placed on the left of this vertical bar, while the constants on the right-hand side of each equation are placed on the right of the vertical bar. The matrices that represent these systems can be manipulated in such a way as to provide easy-to-read solutions. This manipulation is called row reduction. Row reduction techniques transform the matrix into **reduced row echelon form** without changing the solutions to the system. To convert any matrix to its reduced row echelon form, Gauss–Jordan elimination is performed. There are three elementary row operations used to achieve reduced row echelon form [6–63]:

- Switch two rows.
- Multiply a row by any nonzero constant.
- Add a scalar multiple of one row to any other row.

2.5.1.4 LU Decomposition Method

It is a form of Gaussian elimination method. It decomposes the A matrix into the form a lower triangular matrix and an upper triangular matrix. *LU* factorization is a computation intensive method, especially for large matrices [40]. It calculates $A = LU$, L is a lower triangular matrix with unit diagonal entries and U is an upper triangular matrix. The detailed steps of the solution of Eq. (2.31) can be shown as follows:

- Step 1: Let $LUx = b$.
- Step 2: Let $Ux = y$.
- Step 3: Solve $Ly = b$ for y.
- Step 3: Calculate x from step 2.
- Step 4: Solution $= \begin{bmatrix} -1 \\ 3 \\ 2 \end{bmatrix}$.

Another nice feature of the *LU* decomposition is that it can be done by overwriting A, therefore saving memory if the matrix A is very large. The *LU* decomposition is useful when one needs to solve $Ax = b$ for x when A is fixed and there are many different b's. First one determines L and U using Gaussian elimination. Then one writes: $(LU)x = L(Ux) = b$. Then we let: $y = Ux$, and first solve $Ly = b$ for y by forward substitution. We then solve $Ux = y$.

2.5.1.5   Cholesky Decomposition Method

It is a decomposition of a Hermitian, positive-definite matrix into the product of a lower triangular matrix and its conjugate transpose, which is useful for numerical solutions [41, 42]. It is valid only for square and symmetric matrices. So, it cannot be applied to our example in Eq. (2.32).

Cholesky decomposition is a special version of LU decomposition tailored to handle symmetric matrices more efficiently. For a positive symmetric matrix A where $a_{ij} = a_{ji}$, LU decomposition is not efficient enough as computational load can be halved using Cholesky decomposition. In Cholesky, matrix A can be decomposed as follows:

$$A = L * L^T \qquad (2.32)$$

where L is a low triangle matrix.

To compute L:

$$l_{j,j} = \sqrt{a_{j,j} - \sum_{k=0}^{j-1} l_{j,k}^2} \qquad (2.33)$$

$$l_{i,j} = \frac{1}{l_{j,j}} \left(a_{i,j} - \sum_{k=0}^{j-1} l_{i,k} l_{j,k} \right) \text{for } i > j \qquad (2.34)$$

2.5.2   Iterative Methods for Solving SLEs

Iterative methods for solving SLEs can be classified into two categories: stationary methods such as Jacobi, Gauss–Seidel, and Gaussian over-relaxation and Krylov subspace methods such as Conjugate gradient methods which extract the approximate solution from a subspace of dimension much smaller than the size of the coefficient matrix A. This approach is called projection method [26, 34].

The solution of the linear system $Ax = b$ using stationary methods can be expressed in the general form.

$$M X^{n+1} = N X^n + b \qquad (2.35)$$

where $x(k)$ is the approximate solution at the kth iteration. The choice of M and N for different stationary methods are summarized in Table 2.8

For projection methods, let x_0 be an initial guess for this linear system and $r_0 = b - Ax_0$ be its corresponding residual. We continue the iterations until r_0 is very small.

All these methods are discussed in detail in the below subsections.

Table 2.8 Stationary iterative methods for linear systems

Method	M	N
Jacobi	D	L + U
Gauss–Seidel	D−L	U
Gaussian over-relaxation	$\dfrac{1}{\omega}\text{D}-\text{L}$	$\left(\dfrac{1}{\omega}-1\right)\text{D}+\text{U}$

D, L, and U are the diagonal, lower triangular, and upper triangular parts of A, respectively

2.5.2.1 Jacobi Method

Jacobi is an iterative method for solving system of linear equations. The Jacobi method starts with an initial guess of vector x and solves each unknown x. The obtained guess is then used as the current solution and the process is iterated again. This process continues until it converges to a solution [43].

To solve Eq. (2.32) using Jacobi method:

- Step 1: is to arrange Eq. (2.32) as below in an iterative format [44]:

$$x = \frac{8-2y-3z}{4} \tag{2.36}$$

$$y = \frac{-14-3x-2z}{-5} \tag{2.37}$$

$$z = \frac{27+2x-3y}{8} \tag{2.38}$$

- Step 2: substitute in the following equations assuming that x_0, y_0, z_0 are zeros.

$$x_{n+1} = \frac{8-2y_n-3z_n}{4} \tag{2.39}$$

$$y_{n+1} = \frac{-14-3x_n-2z_n}{-5} \tag{2.40}$$

$$z_{n+1} = \frac{27+2x_n-3y_n}{8} \tag{2.41}$$

- Step 3: we continue in iterations until convergence happens. Solution $=\begin{bmatrix} -0.99 \\ 3 \\ 1.99 \end{bmatrix}$ after 20 iterations.

2.5.2.2 Gauss–Seidel Method

The Gauss–Seidel (GS) method, also known as the Liebmann method or the method of successive displacement, is an iterative method used to solve a linear system of equations. It is named after the German mathematicians Carl Friedrich Gauss and Philipp Ludwig von Seidel, and is similar to the Jacobi method. Though it can be applied to any matrix with nonzero elements on the diagonals, convergence is only guaranteed if the matrix is either diagonally dominant, or symmetric and positive definite. It was only mentioned in a private letter from Gauss to his student Gerling in 1823. A publication was not delivered before 1874 by Seidel.

In certain cases, such as when a system of equations is large, iterative methods of solving equations are more advantageous. Elimination methods, such as Gaussian elimination, are prone to large round-off errors for a large set of equations. Iterative methods, such as the Gauss–Seidel method, give the user control of the round-off error. Also, if the physics of the problem are well known, initial guesses needed in iterative methods can be made more judiciously leading to faster convergence.

GS is an improvement of the Jacobi algorithm. We use "new" variable values (subscript $= n + 1$) wherever possible; substitute in the following equations assuming that x_0, y_0, z_0 are zeros and continue in iterations until convergence happens [6, 33].

$$x_{n+1} = \frac{8 - 2y_n - 3z_n}{4} \tag{2.42}$$

$$y_{n+1} = \frac{-14 - 3x_{n+1} - 2z_n}{-5} \tag{2.43}$$

$$z_{n+1} = \frac{27 + 2x_{n+1} - 3y_{n+1}}{8} \tag{2.44}$$

$$\text{Solution} = \begin{bmatrix} -0.99 \\ 3 \\ 1.99 \end{bmatrix} \text{ after 11 iterations.}$$

2.5.2.3 Successive Over-Relaxation (SOR) Method

It combines the parallel nature of Jacobi iterative method used with higher convergence rate. It is an enhanced version of Gauss–Seidel method. We use a relaxation variable ω where generally $1 < \omega < 2$. Notice that if $\omega = 1$, then this is the Gauss–Seidel method. The following equations assuming that x_0, y_0, z_0 are zeros and continue in iterations until convergence happens.

$$x_{n+1} = \omega \frac{8 - 2y_n - 3z_n}{4} \tag{2.45}$$

$$y_{n+1} = \omega \frac{-14 - 3x_{n+1} - 2z_n}{-5} \tag{2.46}$$

$$z_{n+1} = \omega \frac{27 + 2x_{n+1} - 3y_{n+1}}{8} \tag{2.47}$$

Solution to Eq. (2.32) = $\begin{bmatrix} -0.99 \\ 3 \\ 1.99 \end{bmatrix}$ after 4 iterations.

It is often very difficult to estimate the optimal relaxation factor, which is a key parameter of the SOR method.

2.5.2.4 Conjugate Gradient Method

Basic iterative methods such as Jacobi Method and Gauss–Seidel method cannot solve all the linear systems. The Conjugate Gradient method is one of the Krylov subspace methods. The conjugate gradient (GC) method derives its name from the fact that it generates a sequence of conjugate (or orthogonal) vectors. These vectors are the residuals of the iterations. They are also the gradients of a quadratic functional, the minimization of which is equivalent to solving the linear system.

Conjugate Gradient is an Iterative Method applicable to sparse systems that are too large to be handled by a direct implementation. It has the advantage that it reaches the required tolerance after a relatively small number of iterations compared to Jacobi and Gauss methods.

The conjugate gradient method is used to solve equations where the matrix is symmetric [43]. To solve Eq. (2.31) keep it in the form of $Ax = b$, then apply the following algorithm. It is valid only for square and symmetric matrices only. So, it cannot be applied to our example in Eq. (2.31).

The algorithm works as follows:

$$r_0 = b - Ax_0$$

$$p_0 = r_0$$

$$k = 0$$

Repeat

$$\alpha_n = \frac{r_n^T r_n}{p_n^T A p_n}$$

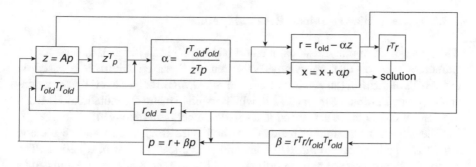

Fig. 2.13 Data-flow diagram for conjugate gradient method

$$x_{n+1} = x_n + \alpha_n P_n$$

$$r_{n+1} = r_n - \alpha_n A P_n$$

If r_{n+1} is very small, then exit the loop, else

$$P_{n+1} = r_{n+1} + \beta_n P_n$$

$$n = n+1$$

End repeat
The solution is x_{n+1}
Data-flow diagram for Conjugate Gradient method is shown in Fig. 2.13.

2.5.2.5   Bi-conjugate Gradient Method

Bi-conjugate Gradient method can solve any linear system. The Bi-conjugate Gradient method generates two CG-like sequences of vectors, one based on a system with the original coefficient matrix, and one on its transposal. Instead of orthogonalizing each sequence, they are made mutually orthogonal, or "bi-orthogonal." This method, like CG, uses limited storage. It is useful when the matrix is nonsymmetric and nonsingular. A singular matrix is a square matrix that has no inverse. A determinant equal zero means that the matrix is singular. It is valid only for square and symmetric matrices only. So, it cannot be applied to our example in Eq. (2.1).

2.5.2.6 Generalized Minimal Residual Method

The generalized minimal residual method (**GMRES**) is an iterative method for the numerical solution of a nonsymmetric system of linear equations. It is based on **Arnoldi's algorithm** (an eigenvalue algorithm). GMRES solves at every step a minimum squares problems (*min (Ax-b)*), which requires many computations. GMRES is useful if convergence is reached in a small number of iterations [45].

The generalized minimal residual method (GMRES) is an iterative method used to find numerical solutions to nonsymmetric linear systems of equations. The method relies on constructing an orthonormal basis of the Krylov space and is thus vulnerable to an imperfect basis caused by computational errors. There have been attempts to address this issue by devising variations of the method that are less sensitive to poorly conditioned problems. The GMRES algorithm is typically used when the dimensions of the problem are very large; thus, it is of interest to investigate ways in which the computational and memory cost of running it can be reduced.

*   **Krylov subspace**: The order -r Krylov subspace generated by an n-by-n matrix A and a vector *b* of dimension n is the linear subspace spanned by the images of b under the first r powers of A (starting from $A^0 = I$), that is, $K_r(A, b) = span \{b, Ab, A^2b, ..., A^{r-1}b\}$. We use Krylov subspace in iterative methods for finding one or a few eigenvalues of large sparse matrices or solving large systems of linear equations avoid matrix-matrix operations, but rather multiply vectors by the matrix and work with the resulting vectors. Starting with a vector, *b*, one computes $A*b$, then one multiplies that vector by A to find A^2b and so on. Because the vectors usually soon become almost linearly dependent due to the properties of power iteration, methods relying on Krylov subspace frequently involve some orthogonalization scheme such as Arnoldi's iteration for more general matrices.
*   **Arnoldi's iteration**: It is an eigenvalue algorithm, also an example of an iterative method and is an analogy for computing the QR factorization $A = QR$ of a matrix A. Arnoldi's iteration finds an approximation to the eigenvalues and eigenvectors of general matrices by forming an orthonormal basis of the Krylov subspace via Gram-Schmidt Orthogonalization, which is useful when dealing with large sparse matrices.

2.5.3   Hybrid Methods for Solving SLEs

Hybrid methods combine direct and iterative methods to exploit the advantages of both direct and iterative methods. It depends on partitioning the matrix and solving its subsets using direct or iterative methods [46]. In [47], the authors proposed a hybrid approach, where the matrix is partitioned into blocks. Within each block, highly optimized (parallel) conventional solver is used, then the blocks are coupled together using block Jacobi. This allows limiting the block size to the point where the conventional iterative methods no longer scale.

Moreover several combinations of evolutionary computation techniques and classical numerical methods are proposed to solve linear equations.

2.5.4   ML-Based Methods for Solving SLEs

It enhances solving SLEs by using artificial intelligence techniques to accelerate the convergence of iterative numerical methods. The proposed method will be based on genetic algorithm (GA) and neural network (NN) artificial intelligence methods [37, 48]. These methods have the potential to significantly accelerate large-scale computational efforts by accelerating the convergence of iterative numerical methods. GA and NN SLEs solver was able to find all possible sets of solutions that are applicable to any given system of linear equations. Conventional methods always produce a set of solutions for a particular system of linear equations. Note that the population random values should be in the A and b range.

2.5.5   How to Choose a Method for Solving Linear Equations

As stated earlier, there exists a variety of algorithms for solving linear systems (Fig. 2.14). Selecting the best solver algorithm depends on the matrix structure as well as on different trade-offs such as computational complexity, memory bottlenecks, convergence properties, and numerical behavior. Comparison between different methods based on the symmetric 3×3 equations in terms of computation time and number of iterations is shown in Table 2.9.

Direct solvers suffer from slow convergence for large systems of equations. Moreover, it takes more time to converge for sparse linear system as compared to dense linear system. So, iterative methods are preferred in that case.

The simulation results show that the SOR method converges faster than the Jacobi and Gauss–Seidel methods. Jacobi method is the slowest one in terms of speed and the highest one in terms of usage of memory. Conjugate gradient is the fastest among all iterative and direct solvers. But, it has one restriction that the matrix should be symmetric. For direct solver, LU Factorization is the best. The solutions are very accurate for all methods.

The criteria on which we can choose which method to be used is summarized in Table 2.10. The two important criteria to take into consideration when choosing a method for solving linear equations are: convergence rate and the cost of calculating of the method.

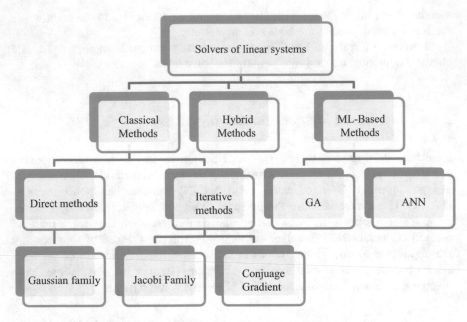

Fig. 2.14 Different methods of solving systems of linear equations

Table 2.9 Comparison between different iterative methods

Methods	Computation time (s)	Number of iterations	Memory usage
Jacobi	0.8	10	The highest
Gauss–Seidel	0.75	9	High
SOR	0.7	8	Average
Conjugate gradient	0.05	6	The lowest
Gaussian elimination	0.03	–	Low
Gauss–Jordan	0.03	–	Low
LU factorization	0.02	–	Low

2.6   Common Hardware Architecture for Different Numerical Solver Methods

CPU and GPU architectures are limited by their maximum memory and computational bandwidth which are considered low compared to FPGA for these types of problems. In this chapter, generalized FPGA-based hardware architecture for different methods used to solve system of linear equations is presented to speed-up the solving time. We build a complete hardware library of most of the famous methods with many shared hardware blocks. The proposed architecture consists of control

Table 2.10 How to choose a method for solving linear equations

		Matrix Type				
		Dense			Sparse	
		Square			Symmetric	Not symmetric
SLEs solver methods		Symmetric	Not symmetric	Not Square		
Direct solver	Cramer	✓	✓	✗	✗	
	Gaussian elimination	✓	✓	✓		
	Gauss–Jordan elimination	✓	✓	✓		
	LU decomposition	✓	✓	✓		
	Cholesky decomposition	✓	✗	✗		
Iterative/ indirect solver	Stationary methods	Gauss–Seidel	✗			✓	✓
		Jacobi				✓	✓
		Gaussian over-relaxation				✓	✓
	Krylov methods	Conjugate gradient				✓	✗
		Bi-conjugate gradient			✓	✓	✗
		GMRES				✗	✓

Fig. 2.15 Block diagram describing the general structure of the proposed generic hardware-based numerical analysis solver

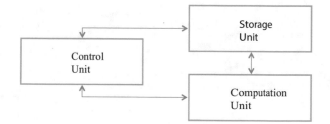

unit, storage unit, and computation unit. Block diagram describing the general structure of the proposed generic hardware-based numerical analysis solver is shown in Fig. 2.15. The proposed methodology is summarized in Fig. 2.16. The proposed parallelized architecture can speed-up the execution time for many applications over software solutions.

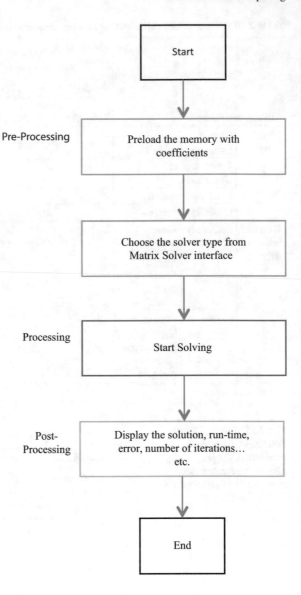

Fig. 2.16 The proposed methodology

2.7   Software Implementation for Different Numerical Solver Methods

2.7.1   Cramer's Rule: Python-Implementation

Cramer's Rule

```python
import numpy as np
def det2x2(A, v=False):
 if v: print 'compute 2 x 2 det of'
 if v: print A
 assert A.shape == (2,2)
 return A[0][0]*A[1] - A[0, 1]*A[0, 1]

def det3x3(A):
 print 'compute 3 x 3 det of'
 print A
 assert A.shape == (3,3)
 a,b,c = A[0]
 c1 = a * det2x2(A[1:3,[1, 2]])
 c2 = b * det2x2(A[1:3,[0,2]])
 c3 = c * det2x2(A[1:3,[0,1]])
 return c1 - c2 + c3

def solve(A):
 print 'solve'
 print A, '\n'
 assert A.shape == (3,4)
 D = det3x3(A[:,:3])
 print 'D = ', D, '\n'
 if D == 0:
 print 'no solution'
 return
 Dx = det3x3(A[:,[1-3]])
 print 'Dx = ', Dx, '\n'
 Dy = det3x3(A[:,[0,3,2]])
 print 'Dy = ', Dy, '\n'
 Dz = det3x3(A[:,[0,1,3]])
 print 'Dz = ', Dz, '\n'
 return Dx*1.0/D, Dy*1.0/D, Dz*1.0/D
```

2.7.2   Newton–Raphson: C-Implementation

Newton Raphson

```
// C++ program for implementation of Newton Raphson Method for
// solving equations
#include<bits/stdc++.h>
#define EPSILON 0.001
using namespace std;
// An example function whose solution is determined using
// Bisection Method. The function is x^3 - x^2 + 2
double func(double x)
{
 return x*x*x - x*x + 2;
}


// Derivative of the above function which is 3*x^x - 2*x
double derivFunc(double x)
{
 return 3*x*x - 2*x;
}


// Function to find the root
void newtonRaphson(double x)
{
 double h = func(x) / derivFunc(x);
 while (abs(h) >= EPSILON)
 {
 h = func(x)/derivFunc(x);

 // x(i+1) = x(i) - f(x) / f'(x)
 x = x - h;
 }

 cout << "The value of the root is : " << x;
}

// Driver program to test above
int main()
```

```
{
 double x0 = -20; // Initial values assumed
 newtonRaphson(x0);
 return 0;
}
```

2.7.3   Gauss Elimination: Python-Implementation

Gauss Elimination

```
def myGauss(m):
 #eliminate columns
 for col in range(len(m[0])):
 for row in range(col+1, len(m)):
 r = [(rowValue * (-(m[row][col] / m[col][col]))) for rowValue in
m[col]]
 m[row] = [sum(pair) for pair in zip(m[row], r)]
 #now backsolve by substitution
 ans = []
 m.reverse() #makes it easier to backsolve
 for sol in range(len(m)):
 if sol == 0:
 ans.append(m[sol][-1] / m[sol][-2])
 else:
 inner = 0
 #substitute in all known coefficients
 for x in range(sol):
 inner += (ans[x]*m[sol][-2-x])
 #the equation is now reduced to ax + b = c form
 #solve with (c - b) / a
 ans.append((m[sol][-1]-inner)/m[sol][-sol-2])
 ans.reverse()
 return ans
```

2.7.4   Conjugate Gradient: MATLAB-Implementation

Conjugate Gradient

```
close all;
clear all;
clc;
prompt = 'Please Enter the A Matrix, taking into consideration A
being symmetric & positive: ';
A = input(prompt)
prompt = 'Pleae Enter The b Vector';
b = input(prompt)
prompt = 'Please Enter Initial guess: ';
x = input(prompt);
r = b-A*x;
p = r;
rcomp = r'*r
 for i = 1:length(b)
 Ap = A * p;
 alpha = rsold / (p' * Ap);
 x = x + alpha * p;
 r = r - alpha * Ap;
 rsnew = r' * r;
 if sqrt(rsnew) < 1e-10
 break;
 end
 p = r + (rsnew / rsold) * p;
 rsold = rsnew;
 end
```

2.7.5   GMRES: MATLAB-Implementation

GMRES

```
%% Initialization
clc
clear all
close all

%% Code
A = [3 4 5; 6 7 8; 9 10 11]; %nxn matrix
b = [1 ;1 ;1]; %nx1 vector
n=size(A,1); %To get its size
x0 = [-4 ;5 ;-9]; %Arbitrary guess
r0 = b - A*x0; %Our guess
k = 5; %Number of iterations,
 %If we increase k we increase
 %accuracy but computational time increases.

V=zeros(n,k+1); %Our n x k+1 matrix
H=zeros(k+1,k); %Our upper matrix.
V(:,1)=r0/norm(r0); %Our initial guess of V.

% Arnoldi, The Gram-Schmidt implementation
for j=1:k
 z=A*V(:,j);
 for i=1:j
 H(i,j)=dot(z,V(:,i)); %Dot product
 z=z-H(i,j)*V(:,i);
 end
 H(j+1,j)=norm(z);
 if H(j+1,j)==0, break, end
 V(:,j+1)=z/H(j+1,j);
end
```

2.7.6   Cholesky: MATLAB-Implementation

Cholesky

```
%% Initialization
clc
clear all
close all

%% Code
A = [3 4 5; 6 7 8; 9 10 11]; %nxn matrix
b = [1 ;1 ;1]; %nx1 vector
n=size(A,1); %To get its size
x0 = [-4 ;5 ;-9]; %Arbitrary guess
r0 = b - A*x0; %Our guess
k = 5; %Number of iterations,
 %If we increase k we increase
 %accuracy but computational time increases.

V=zeros(n,k+1); %Our n x k+1 matrix
H=zeros(k+1,k); %Our upper matrix.
V(:,1)=r0/norm(r0); %Our initial guess of V.

% Arnoldi, The Gram-Schmidt implementation
for j=1:k
 z=A*V(:,j);
 for i=1:j
 H(i,j)=dot(z,V(:,i)); %Dot product
 z=z-H(i,j)*V(:,i);
 end
 H(j+1,j)=norm(z);
 if H(j+1,j)==0, break, end
 V(:,j+1)=z/H(j+1,j);
end
```

2.8 Conclusions

In this chapter, we introduce the numerical analysis methods for electronics. Numerical analysis is the study of algorithms that use numerical approximation for the problems of mathematical analysis. Numerical analysis naturally finds application in all fields of engineering and the physical sciences. Also the life sciences, social sciences, medicine, business, and even the arts have adopted elements of scientific computations.

In this chapter, we introduce different approaches to solve partial differential equations (PDEs) and ordinary differential equations (ODEs) as well as the advantages and disadvantages of each method are analyzed. In this chapter, we introduce different approaches to solve system of nonlinear equations (SNLEs) as well as the advantages and disadvantages of each method are analyzed. In this chapter, we introduce different approaches to solve system of linear equations (SLEs) as well as the advantages and disadvantages of each method are analyzed.

References

1. K. Salah, A novel model order reduction technique based on artificial intelligence. Microelectron. J. **65**, 58–71 (2017)
2. S.K. Gupta, *Numerical Methods for Engineers* (Wiley Eastern, New Delhi, 1995)
3. W.W. Hager, *Applied Numerical Linear Algebra* (Prentice Hall, Englewood Cliffs, 1988)
4. J.J. Dongarra, I.S. Du, D.C. Sorensen, H.A. van der Vorst, *Numerical Linear Algebra for High-Performance Computers* (SIAM, Philadelphia, 1998)
5. J.D. Hoffman, *Numerical Methods for Engineers and Scientists* (McGraw-Hill, Inc., Singapore, 1992)
6. P. Raghavan, K. Teranishi, E.G. Ng, A latency tolerant hybrid sparse solver using incomplete cholesky factorization. Num Lin Alg Appl **10**, 541–560 (2003)
7. S.H. Gerez, *Algorithms for VLSI Design Automation* (Wiley, Chichester, 1999)
8. L.-T. Wang, Y.-W. Chang, K.-T.T. Cheng, *Electronic Design Automation: Synthesis, Verification, and Test* (The Morgan Kaufmann Series in Systems on Silicon, Morgan Kaufmann, 2009)
9. D. Jansen, *The Electronic Design Automation Handbook* (Kluwer Academic, Boston, MA, 2003)
10. C.J. Alpert, D.P. Mehta, S.S. Sapatnekar, *Handbook of Algorithms for Physical Design Automation* (CRC, Boca Raton, 2009)
11. A.B. Kahng, J. Lienig, I.L. Markov, J. Hu, *VLSI Physical Design: From Graph Partitioning to Timing Closure* (Springer Science+Business Media, New York, 2011)
12. S.M. Sait, H. Youssef, *VLSI Physical Design Automation: Theory and Practice, Volume 6 of Lecture Notes Series on Computing* (World Scientific Publishing, Singapore, 1999)
13. M. Sarrafzadeh, C.K. Wong, *An Introduction to VLSI Physical Design*, McGraw-Hill Series in Computer Science (McGraw-Hill, New York, 1996)
14. N.A. Sherwani, *Algorithms for VLSI Physical Design Automation*, 3rd edn. (Kluwer Academic, Norwell, 1999)
15. https://en.wikipedia.org/wiki/Finite_difference_method
16. G. Strang, G. Fix, *An Analysis of the Finite Element Method* (Prentice Hall, Englewood Cliffs, 1973)

17. A.C. Polycarpou, *Introduction to the Finite Element Method in Electromagnetics* (Morgan & Claypool, San Rafael, 2006)
18. http://mathworld.wolfram.com/LegendrePolynomial.html
19. http://mathworld.wolfram.com/Fourier-LegendreSeries.html
20. G.M. Diaz-Toca, L. GonzalezVega, H. Lombardi, Generalizing Cramers rule: Solving uniformly linear Systems of Equations. SIAM J Matr Anal Appl **27**, 621–637 (2005)
21. A. Hussian, Numerical solution of partial differential equations by using modified artificial neural network. Net Compl Sys **5**, 6 (2015)
22. J. He, J. Xu, X. Yao, Solving equations by hybrid evolutionary computation techniques. IEEE Trans. Evol. Comput. **4**, 3 (2000)
23. S. Shahzadehfazeli, An improved iterative method for solving general system of equations via genetic algorithms. Int J Inform Technol Model Comp **4**, 1 (2016)
24. http://www.ugrad.math.ubc.ca/coursedoc/math100/notes/approx/newton.html
25. https://www.ijser.org/researchpaper/Newton-Raphson-Method.pdf
26. http://www.seas.ucla.edu/~vandenbe/236C/lectures/qnewton.pdf
27. https://www.stat.cmu.edu/~ryantibs/convexopt/lectures/quasi-newton.pdf
28. The official University of British Columbia Math Division website: https://www.math.ubc.ca/~pwalls/math-python/roots-optimization/secant/
29. The open-source java codes website: http://theflyingkeyboard.net/java/java-secant-method-2/
30. https://www.geeksforgeeks.org/program-muller-method/
31. N. Flyer, A hybrid analytical–numerical method for solving evolution partial differential equations. I. The half-line. Proceeding of the Royal Society (2008)
32. A. Dhamacharon, An efficient hybrid method for solving systems of nonlinear equations. J. Comput. Appl. Math. **263**, 59–68 (2014)
33. Y. ZhongLuo, Hybrid approach for solving systems of nonlinear equations using chaos optimization and quasi-Newton method. Appl. Soft Comput. **8**(2), 1068–1073 (2008)
34. H. Fathalizadeh, *Solving Nonlinear Ordinary Differential Equations Using Neural Networks*. 2016 4th International Conference on Control, Instrumentation, and Automation (ICCIA) (2016), pp. 27–28
35. G. Joshi, *Solving System of Non-Linear Equations using Genetic Algorithm*, International Conference on Advances in Computing, Communications and Informatics (ICACCI) (2014)
36. K. Habgood, I. Arel, A condensation-based application of Cramers rule for solving large-scale linear systems. J Discr Algor (2012)
37. B. Hochet, P. Quinton, Y. Robert, Systolic Gaussian elimination overGF(p) with partial pivoting. IEEE Trans. Comput. **38**(9), 1321–1324 (1989)
38. http://caslab.csl.yale.edu/code/gausselim/
39. I. Kyrchei, Cramer's rule for quaternionic systems of linear equations. J. Math. Sci. **155**, 839–858 (2012)
40. W.H. Press, S.A. Teukolsky, W.T. Vetterling, B.P. Flannery, *Numerical Recipes (the Art of Scientific Computing)*, 3rd edn. (Cambridge University Press, Cambridge, 2007)
41. R.L. Burden, J.D. Faires, *Numerical Analysis*, 8th edn. (Thomson Learning, Boston, 2005)
42. S. Moussa, A.M.A. Razik, A.O. Dahmane, H. Hamam, *FPGA Implementation of Floating-Point Complex Matrix Inversion Based on GAUSS-JORDAN Elimination*. In 26th Annual IEEE Canadian Conference on Electrical and Computer Engineering (CCECE) (Regina, Canada 2013)
43. R. Duarte, H. Neto, M. Vestias, *Double-Precision Gauss-Jordan Algorithm with Partial Pivoting on FPGAs*. In 12th Euromicro Conference on Digital System Design, Architectures, Methods and Tools (DSD) (2009), p. 273–280
44. https://www.maa.org/press/periodicals/loci/joma/iterative-methods-for-solving-iaxi-ibi-jacobis-method.
45. D. Yang, et al, *Performance Comparison of Cholesky Decomposition on GPUs and FPGAs*, In: Symposium on Application Accelerators in High Performance Computing (2010)
46. https://en.wikipedia.org/wiki/Krylov_subspace

47. https://en.wikipedia.org/wiki/Arnoldi_iteration
48. G.R. Morris, K.H. Abed, Mapping a Jacobi iterative solver onto a high performance heterogeneous computer. IEEE Trans Parall Distrib Sys **24**(1), 85–91 (2013)
49. G. Wu, Y. Dou, J. Sun, G.D. Peterson, A high performance and memory efficient LU decomposer on FPGAs. IEEE Trans. Comput. **61**(3), 366–378 (2012)
50. D. Yang, G.D. Peterson, H. Li, *High Performance Reconfigurable Computing for Cholesky Decomposition*, In Symposium on Application Accelerators in High Performance Computing (SAAHPC), UIUC (2009)
51. N. Brown, *Solving Large Sparse Linear Systems using Asynchronous Multi-splitting*. PRACE (2013)
52. C.F. Gerald, P.O. Wheatley, *Applied Numerical Analysis*, 5th edn. (Addison-Wesley, Singapore, 1998)
53. C. Kelley Iterative methods for linear and nonlinear equations. Society for Industrial Mathematics (1995)
54. K. Salah, A generic model order reduction technique based on Particle Swarm Optimization (PSO) algorithm. IEEE EUROCON (2017)
55. http://mathforcollege.com/nm/mws/gen/04sle/mws_gen_sle_txt_seidel.pdf
56. http://www.preyeshdalmia.com/assets/Iterative_solver.pdf
57. https://www.sanfoundry.com/cpp-program-implement-gauss-seidel-method/
58. https://github.com/nuhferjc/gauss-seidel/blob/master/gaussseidel.py
59. https://matrix.reshish.com/cramer.php
60. https://perso.uclouvain.be/paul.vandooren/Krylov.pdf
61. http://caslab.csl.yale.edu/publications/2016_reconfig.pdf
62. https://www.austincc.edu/jthom/GaussJordanElimination.pdf
63. https://www.codewithc.com/c-program-for-gauss-jordan-method/

Chapter 3
Parallel Computing: OpenMP, MPI, and CUDA

3.1 Introduction

Golden Moore predicted that the number of transistors in an integrated circuit doubles every 18 months. This law is known as Technology Growth Law. Now, Moore's law reaches saturation due to manufacturing, materials, and design limitations. New trends have been proposed to evade Moore's law. Here we will present the parallel computing. Parallel computing/programming is a computer programming technique that enables parallel execution of operations. It uses multiple processors in parallel to solve problems more quickly than with a single processor. If you cannot increase the clock, do more operations by one clock. But, we cannot build an infinite processor due to temperature, cooling problems, interconnect bottleneck, etc. [1–5]. Performance gained by multicore processor strongly dependent on the software algorithms and implementation. Moore's law continues to provide more transistors, but threshold, and thus supply voltages no longer scale down each generation. Hence, to leverage the additional transistors, designers reduce clock frequency and invest the power head-room to activate twice as many cores. Because of the quadratic relationship between power and voltage, a small voltage/frequency reduction allows a doubling of the number of transistors that switch each cycle, enabling overall throughput improvements from multicore. We cannot just reduce the size of transistors to improve performance as smaller transistors → faster processors → increased power consumption → increased heat → unreliable processors. Parallel computing applications include:

- *Movie making* Many days are needed to simulate just few minutes.
- *EM Simulation* Many days are needed to simulate 3D structures.
- *Oil Exploration* Months of processing data.
- *Atmosphere Simulation [weather forecasting* Due to global warming, we need to simulate how the weather will be in the next 10 years.
- *Biological* How genes evolve?
- *Defense* Cryptology.

© Springer Nature Switzerland AG 2020
K. S. Mohamed, *Neuromorphic Computing and Beyond*,
https://doi.org/10.1007/978-3-030-37224-8_3

Fig. 3.1 Sequential
example

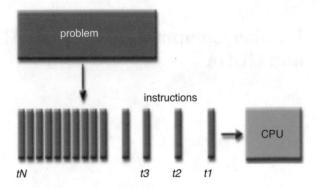

- *Matrix multiplication* can be done using parallel programming as some operations are independent of each other.

Supercomputer refers to a general-purpose computer that can solve computational intensive problems faster than traditional computers. A Supercomputer **may or may not** be a parallel computer. Supercomputers speed are measured in floating point operations per second (FLOPS). Parallel is a subset of super.

3.1.1 Concepts

- **Algorithm**: The sequence of steps necessary to do a computation.
- **Sequential**: Must occur in a strict order such as sun-set then sun-rise (Fig. 3.1).
- **Parallel/Concurrent**: Many events happen simultaneously such as music/song, home construction, and call center (Fig. 3.2).
- **Pipelining**: A form of parallelism, like an assembly line in a factory (Car Factory). It can start the next independent operation before the previous result is available (Fig. 3.3).
- **Thread**: sequence of instructions. It is subroutine in the main program (Fig. 3.4).
- **Latency**: Time for one task to pass from start to end or execution time for a single task. It is always generally used with the number of clock cycles it takes from the data entering a block to the time it leaves the block.
- **Delay**: is generally the propagation delay of a signal through combinational logic. Its units are picoseconds or nanoseconds. Logic designers and architects worry about latency. Circuit designers and physical designers worry about delay.
- **Throughput:** How many data I get during a fixed period of time? Do you keep feeding me with continuous data?
- **Scalability:** When the system increases in size, the performance (efficiency) should also increase.
- **Benchmark:** A *benchmark* is "a standard of measurement or evaluation." Running the same computer benchmark on multiple computers allows a comparison to be made. A computer benchmark is typically a **computer program** that performs a strictly defined set of operations—a *workload*. **Returns** some form of result—a *metric*—describing how the tested computer performed.

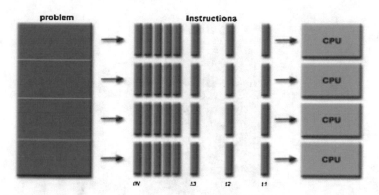

Fig. 3.2 Parallel example

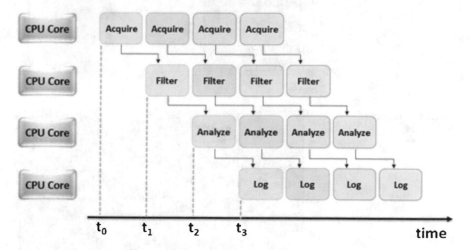

Fig. 3.3 Pipelining example

3.1.2 Category of Processors: Flynn's Taxonomy/Classification (1966)

We have two axes, horizontal for data, vertical for instruction. We have four categories for combinations of data/instruction as shown in Fig. 3.5. SIMD has many execution units, so it can work on different data. More details are shown in Fig. 3.6. Programmability versus efficiency for different classifications are shown in Fig. 3.7 [6].

3.1.2.1 Von-Neumann Architecture (SISD)

John Von-Neumann first authored the general requirements for an electronic computer in 1945. Data and instructions are stored in memory. Control unit fetches instructions/data from memory, decodes the instructions, and then sequentially

Fig. 3.4 Threading
example

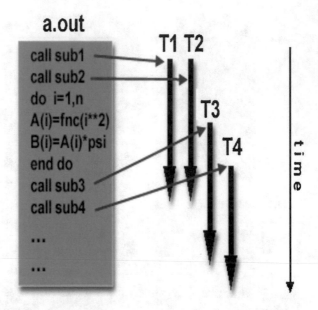

Fig. 3.5 Category of
processors

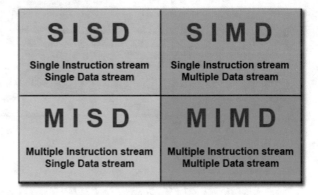

coordinates operations to accomplish the programmed task. Arithmetic unit performs basic arithmetic operations. Input/Output is the interface to the human operator. SISD is shown in Fig. 3.8.

3.1.2.2 SIMD

It is a type of parallel computer. All processing units execute the same instruction at
any given clock cycle. Each processing unit can operate on a different data element.
All ALUs are required to execute the same instruction, or remain idle. Figure 3.9
shows an example for SIMD.

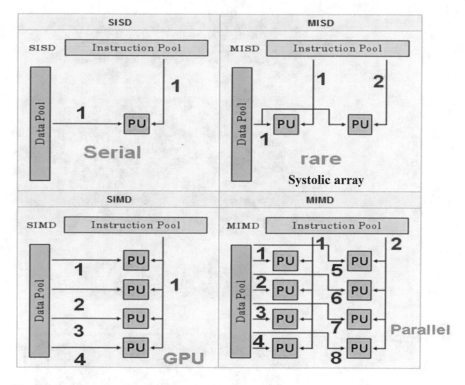

Fig. 3.6 Category of processors: more details

Fig. 3.7 Programmability
versus efficiency

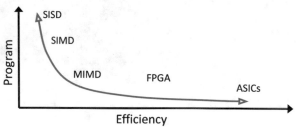

3.1.2.3 MISD

A single data stream is fed into multiple processing units. Each processing unit
operates on the data independently via independent instruction streams. It is generally used for fault tolerance. Heterogeneous systems operate on the same data
stream and must agree on the result. Figure 3.10 shows MISD example. **Systolic**
arrays are example for MISD.

Fig. 3.8 SISD

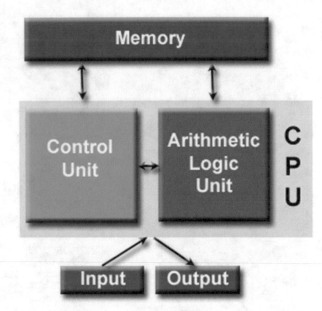

Fig. 3.9 SIMD example

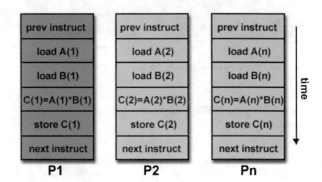

3.1.2.4 MIMD

Currently, it is the most common type of parallel computer. Every processor may be executing a different instruction stream. Every processor may be working with a different data stream. Figure 3.11 shows MIMD example. In an SIMD machine, all processors execute the same instruction at the same time. Hence it is easy to implement synchronization in hardware. In an MIMD machine, different processors may execute different instructions at the same time and it is difficult to support synchronization in hardware. It is done in software.

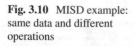

Fig. 3.10 MISD example: same data and different operations

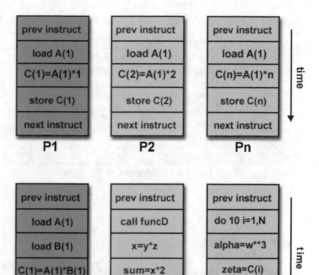

Fig. 3.11 MIMD example

3.1.3 Category of Processors: Soft/Hard/Firm

A soft-core processor is a microprocessor fully described in software, usually in an HDL, which can be synthesized in programmable hardware, such as FPGAs. MicroBlaze soft processor from Xilinx is an example. Firm processor has already been partially synthesized for the particular hardware platform at gate level. Hard processor is implemented in ASIC. Comparison between soft and hard processors is shown in Table 3.1 [11].

3.1.4 Memory: Shared-Memory vs. Distributed Memory

The shared memory is seen by all the processors as a global address space. Processor "n" puts data in the shared memory, the other processors can see it. Locks/semaphores to control access to shared memory. OpenMP Programming is used for shared memory (multicores in one processor). Figure 3.12 shows an example for shared memory.

In distributed memory, each processor is connected to its own memory (invisible to other Processors), it sends message to share data. MPI Programming is used for distributed memory (multiprocessors/multi-computers). Figure 3.13 shows an example for distributed memory.

Table 3.1 Comparison between soft and hard processors

	Soft processors	Hard processors
Definition	A soft processor is a microprocessor core that can be wholly be implemented using logic synthesis. It can be implemented via different semiconductor devices containing programmable logic (e.g., ASIC, FPGA, CPLD).	A hard-core processor is a processor that is actually physically implemented as a structure in the silicon.
Configurability	High.	Low.
Predictability	Low.	High.
Speed	Limited by the speed of the fabric (250 MHz).	Can achieve much faster processing speeds (100's of MHz up to 1+ GHz) since they are optimized and not limited by fabric speed.
Power consumption	High.	Low.
Cost & resources consumption	Implementing a CPU in FPGA fabric is very resource intensive, particularly if you want a lot of computing power.	Cheaper.
Area	Takes big area as FPGA has many components and not all of them used in all application, but its configurability can save some of the missing area.	Fixed area depending on the application.
Others	• Can be easily modified and tuned to specific requirements, more features, custom instructions, etc. • Multiple cores can be used (at the cost of resources). • Can be reconfigured at runtime.	Fixed and cannot be changed.
Examples	PULPino [7], LEON3 [8], MicroBlaze [9], and Nios II [10].	ARM

Fig. 3.12 Shared memory

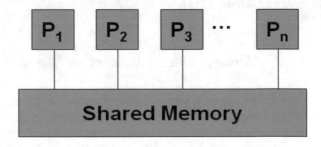

The shared memory can be uniform memory access (UMA) or nonuniform memory access (NUMA). In UMA (Fig. 3.14), we have identical processors, equal access and access times to memory, and the OS treats them the same way. In NUMA (Fig. 3.15), not all processors have equal access to all memories and memory is near from some processors, far from other processors [12].

Fig. 3.13 Distributed memory

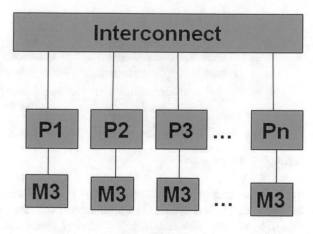

Fig. 3.14 UMA

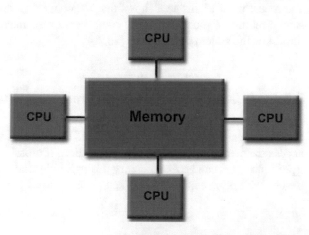

Fig. 3.15 NUMA

Sometimes, we have a trade-off between programmability and scalability, so we use a hybrid solution as shared memory is easier in programming and distributed memory is easier in scalability as shown in Fig. 3.16.

In order to utilize the full potential of the hardware platform, a many-core OS must be aware of the topology and typical memory-access latencies. Based on that information, the placement of processes and data must be optimized at runtime. The cache coherence protocols, which make sure that a memory modification by one core does not get unnoticed by the others, significantly contribute to the traffic on the on-chip interconnect and, thus, has a negative impact on the performance. It has been shown

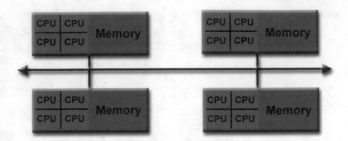

Fig. 3.16 Hybrid solution

that software in general and especially the OS can influence this traffic. Traditionally, the focus of NUMA-aware OS optimizations was set on avoiding transfer delays caused by memory distance. However, modern systems have quite fast interconnects, and recent work in this area shows that the congestion due to concurrently accessed shared resources, such as memory controllers or the interconnect itself, has a bigger impact on the system performance [13, 14].

3.1.5 Interconnects: Between Processors and Memory

Interconnects can be buses or switches. If buses are used, only one processor can access the memory at a time using arbitration. If crossbar switch is used, then multiple processors can access memory through independent paths and it is faster than buses. Arbitration can be static (fixed priority) or dynamic (TDMA).

3.1.6 Parallel Computing: Pros and Cons

Parallel computing/programming is a computer programming technique that enables parallel execution of operations by using multiple processors in parallel to solve problems more quickly than with a single processor.

- **Pros**

    - Performance.
    - Cost-effectiveness.

- **Cons**

    - Software.
    - It is difficult to make many processor work together efficiently.
    - It is difficult to write programs to utilize multiple processors at once in an efficient manner.

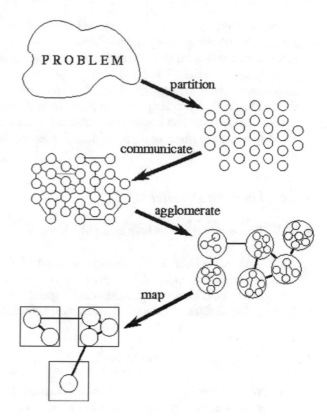

Fig. 3.17 Typical steps for constructing a parallel algorithm

3.2 Parallel Computing: Programming

If you run your software on a single-core computer only, parallel programming is not worth the overhead. The parallel programming is harder than serial one. Running serial code using only one thread parallel programming will take longer time due to parallelism overhead. You should always use more than one thread.

3.2.1 Typical Steps for Constructing a Parallel Algorithm

The typical steps to construct a parallel program (Fig. 3.17):

- Identify what pieces of work can be performed concurrently.
- Partition concurrent work onto independent processors. The mapping of work to processors can be done statically by the programmer or dynamically by the runtime.

- Distribute a program's input, output, and intermediate data: If data is in shared memory, distributing it may be unnecessary.
- Coordinate accesses to shared data: avoid conflicts: Ensure proper order of work using synchronization (Fig. 3.18).
- *Agglomeration:* is an optional task. The task and communication structures defined in the first two stages of a design are evaluated with respect to performance requirements and implementation costs. If necessary, tasks are combined into larger tasks to improve performance or to reduce development costs.

3.2.2   Levels of Parallelism

3.2.2.1   Processor: Architecture Point of View

- Arithmetic Parallelism: The arithmetic operations ($+ - * /$) are each done in a separate execution unit. This allows several execution units to be used simultaneously, because the execution units operate independently.
- Memory Parallelism: Memory is divided into multiple banks.

3.2.2.2   Programmer Point of View

- Task/Program Partitioning: How to split a single task among the processors so that each processor performs the same amount of work, and all processors work collectively to complete the task.
- Data Partitioning: How to split the data evenly among the processors in such a way that processor interaction is minimized.

Fig. 3.18 Synchronization overhead

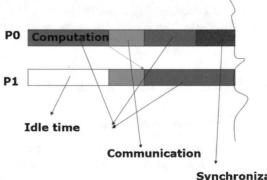

3.3   Open Specifications for Multiprocessing (OpenMP) for Shared Memory

It is an Application Programming Interface (API) for explicit multithreaded, shared memory parallelism. If OpenMP is disabled when compiling a program, the program will execute sequentially. **It is an extension of C, C++, and FORTRAN.** It has a long history, standard defined by a consortium [15]:

– Version 1.0, released in 1997.
– Version 2.5, released in 2005.
– Version 3.0, released in 2008.
– Version 3.1, released in 2011.
– Version 4.5, released in 2015.

We get speedup by running multiple threads simultaneously. Sequential code is the master thread. OpenMP Execution Model is shown in Fig. 3.19.

OpenMP has some constructs to support synchronization. Each task performs its work until it reaches the barrier. It then stops, or "blocks." Mutex: Only one task at a time may use (own) the lock. Without specifying how to share work, all threads will redundantly execute all the work (i.e., no speedup!). Elements of shared-memory programming are summarized in Fig. 3.20. Sequential code versus OpenMP code example is shown in Fig. 3.21. OpenMP does not check for data dependencies, data conflicts, deadlocks, or race conditions. You are responsible for avoiding that yourself. OpenMP is not something that you install. It comes with your compiler. You just need a compiler that supports OpenMP and you need to know how to enable OpenMP support since it is usually disabled by default. The standard compiler for Windows comes from Microsoft and it is the Microsoft Visual C/C++ compiler from Visual Studio. There are other compilers available as well, both Intel C/C++ Compiler (commercial license required) and GCC (such as MinGW, freely available). The most used OpenMP clauses are shown in Table 3.2.

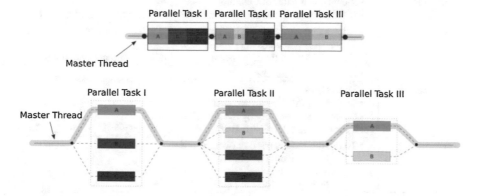

Fig. 3.19 OpenMP Execution Model

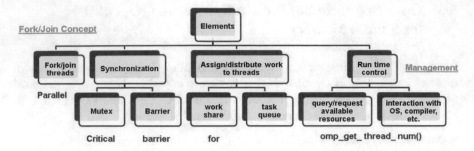

Fig. 3.20 Elements of shared-memory programming

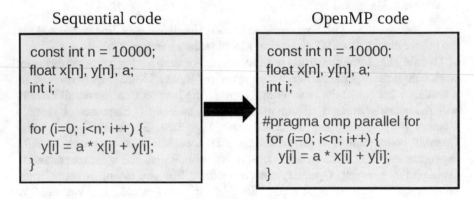

Fig. 3.21 Sequential code versus OpenMP code

Table 3.2 OpenMP clauses

Clause	Explanation
num_threads	• Independent paths of execution • If I did not set it, the compiler will use the **default number** of threads.
nowait	• Allows threads that finish earlier to proceed without waiting • This is useful inside a big parallel region • Example: search for data and when I find it, exit as no need to continue
if	• Conditional statement
Private/shared	• **Data types** • Shared: all threads can access it. • Private: accessed only by the threads that owns it. #pragma omp parallel private (var2,var2) shared (var3)

For	• To run the loop in parallel (work sharing)
	      #pragma omp **parallel**
	      #pragma omp **for**
	      For (i=0;i<10;i++) { }

Fig. 3.22 MPI

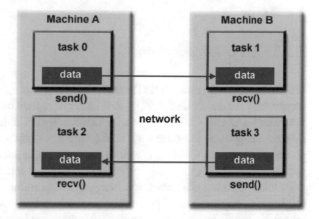

Fig. 3.23 MPI: Abstract example

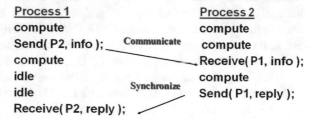

3.4   Message-Passing Interface (MPI) for Distributed Memory

It is a method of programming parallel computers, where the user makes calls to libraries to explicitly share information between processors.

- MPI is a library specification.
- MPI is not a language.
- MPI is a library of **function calls**.

  It has a long history [16]:

- Version 1.0, released in 1994.
- Version 2.0, released in 1996.
- Version 3.0, released in 2012.
- Version 3.1, released in 2015.

Data is exchanged through sending and receiving messages as shown in Fig. 3.22. Abstract example is shown in Fig. 3.23. All processors execute the same program. All processors initiate MPI and calculate their part of the result. Processors 1 to N-1 send result to processor 0. Processor 0 receives results from processors 1 to N-1. Processor 0 prints the final result. All processors exit MPI.

3.5 GPU

3.5.1 GPU Introduction

Graphics processing unit (GPU) consists of hundreds of cores vs. 2, 4, 8, 32 cores on central processing unit (CPU), so it increases speed by 10×, 100×. A comparison between GPU and CPU is shown in Table 3.3. GPUs and CPUs are designed for different purposes. They are both able to provide significant advantages when used at the right place. CPUs are optimized for sequential code performance. Their sophistic control logics allow instructions from a single thread execution to run in parallel. It also has the power to change the sequential order while globally maintaining the appearance of sequential execution. Big cache memories reduce the instruction and data access latencies of large complex applications. On one hand, GPUs' memory interfaces show higher latency, which is readily hidden by massive parallel execution. But on the other hand, they provide a large bandwidth, which is usually many times higher than on CPUs (Up to 150 GB/s). As stated earlier, GPUs are also capable of performing huge numbers of floating-point calculations within short times. CPU architecture versus GPU is shown in Fig. 3.24.

Today's microprocessors are mainly following two different lines: the multicore and the many-core architecture. The multicore architecture is focusing on enhancing sequential tasks with its multiple processor cores. Starting with 2 cores, the

Table 3.3 Comparison between GPU and CPU

	CPU	GPU
Cache	Huge	Small (1)
ALU	Small	Huge
Example	Intel (Xeon)	Nvidia (Tesla)
Optimize	Latency	Throughput
# Threads	Physical cores	Amount of data

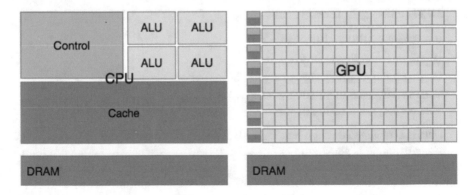

Fig. 3.24 CPU versus GPU

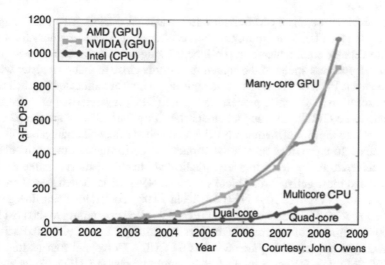

Fig. 3.25 Performance GAP CPU vs. GPU

number of cores is approximately doubling each design generation. The cores themselves come with a full instruction set. Many-core architectures are focusing on the execution throughput of parallel applications. They provide large number of cores each of which is heavily multithreaded. The many-core architecture has seen massive improvements considering performance over CPUs over the last years. Figure 3.25 shows how GPUs are beating CPUs in terms of Floating Point operations per second (FLOPS). Regarding this fact, developers are starting to put computationally intensive parts of their software to the GPU. These high-performance tasks often also require high parallelism [17].

3.5.2 GPGPU

The modern 3D graphics processing unit (GPU) has evolved from a fixed-function graphics pipeline to a programmable parallel processor with computing power exceeding that of multicore.

CPUs. In November 2006, NVIDIA Corporation introduced Tesla architecture which unifies the vertex and pixel processors and extends them to enable high-performance parallel computing applications. The basic computing unit of GPGPU is called streaming multiprocessors (SM).

In March 2010, NVIDIA Corporation introduced Fermi architecture, and GF100 with Fermi architecture has 4 graphics processing clusters (GPC), 16 SMs and 512 cores. For Fermi architecture, each SM has 32 cores, 12 KB L1 cache and 2-warps scheduling. In May 2012, NVIDIA Corporation introduced Kepler architecture. For

Kepler architecture, each SM contains 192 scalar processors (SP) and 32 special function units (SFU). In addition, each SM contains 64K shared memory for threads to share data or communicate in the block. Using the model explicitly to access memory, the access speed of the shared memory is close to that of register without bank conflict. Each SM contains some registers, which are allocated by each thread in the execution. A graphics processing cluster (GPC) is composed of 2 SMs. Two SMs share one GPC and L1 and texture cache. Only four GPCs share the L2 cache. All SMs share the global memory. NVIDIA launched Maxwell architecture in 2014. This architecture provides substantial application performance improvements over prior architectures by featuring large dedicated shared memory, shared memory atomics, and more active thread blocks per SM. NVIDIA launched Pascal architecture in 2016. NVIDIA's new NVIDIA Tesla P100 accelerator using the ground-breaking new NVIDIA Pascal GP100 GPU takes GPU computing to the next level. GP100 is composed of an array of Graphics Processing Clusters (GPCs). Each GPC inside GP100 has ten SMs. Each SM has 64 CUDA Cores and four texture units. With 60 SMs, GP100 has a total of 3840 single precision CUDA Cores and 240 texture units. Tesla P100 features NVIDIA's new high-speed interface, NVLink, which provides GPU-to-GPU data transfers at up to 160 Gb/s of bidirectional bandwidth which is five times the bandwidth of PCIe Gen 3×16.

GPGPU computing is a General-Purpose GPU computing. The GPU can be used for more than just graphics: the computational resources are there, and they are most of the time underutilized. GPU can be used to accelerate data parallel parts of an application. It has simple architecture optimized for compute intensive task. GPGPU is the idea of using a GPU to do the type of computing tasks normally done by the more generalized CPU in a computer. It has large number of cores and low power consumption (green computing). The GPU hardware evolution thus far has gone from an extremely specific, single core, fixed function hardware pipeline implementation just for graphics rendering to a set of highly parallel and programmable cores for more general-purpose computation.

3.5.3 GPU Programming

There are many languages that can be used to program GPUs. They are summarized in Fig. 3.26.

3.5.3.1 CUDA

Compute Unified Device Architecture (CUDA) is used for programming GPUs. CUDA is C-like. CUDA is written in C with extensions to express parallelism. Used for shared-memory. Introduced on 2006. Current version is v7.5 @2015 [18]. GPUs are used for both graphics and non-graphic processing applications. For CUDA, a parallel system consists of a **host** (i.e., CPU) and a computation resource or **device**

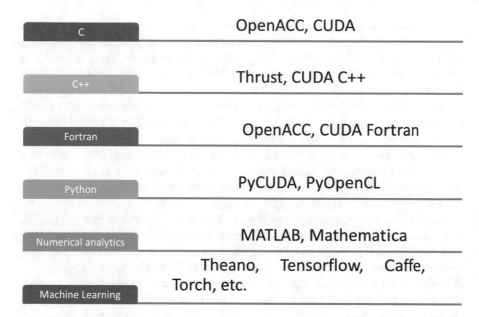

Fig. 3.26 GPU Programming

(i.e., GPU). The computation of tasks is done in the GPU by a set of threads running in parallel. In heterogonous systems, we have CPUs and GPUs which together can process many elements, CPUs with several cores, each core is operating on one or two threads in any given moment, whereas with GPUs hundreds of cores, each core is working on tens or hundreds of threads simultaneously, so in overall, there is thousands of threads that the GPU is working at any given time. The CUDA platform is designed to work with programming languages such as C, C++, and FORTRAN. It allows software developers to use GPUs for general-purpose processing—an approach known as GPGPU [19].

A GPU is built around an array of Streaming Multiprocessors (SMs). A multithreaded program is partitioned into blocks of threads that execute independently from each other, so that a GPU with more multiprocessors will automatically execute the program in less time than a GPU with fewer multiprocessors. Blocks are organized into a one-dimensional, two-dimensional, or three-dimensional grid of threads. There is a limit to the number of threads per block, since all threads of a block are expected to reside on the same processor core and must share the limited memory resources of that core. On current GPUs, a thread block may contain up to 1024 threads. A group of 32 threads are called a warp. Threads in a warp execute in SIMD fashion.

CUDA also enables automatic scalability. Indeed, each block of threads can be scheduled on any of the available multiprocessors within a GPU, in any order, concurrently or sequentially, so that a compiled CUDA program can execute on any number of multiprocessors and only the runtime system needs to know the physical multiprocessor count.

CUDA threads make use of five memories in a GPU. The global memory which is shared among all threads, the shared memory which can be accessed by threads of the same block then each thread has its private local memory. There are also two additional read-only memory spaces accessible by all threads (the constant and texture memory spaces). Global memory bandwidth is most efficient when the simultaneous memory accesses by threads in a half-warp can be coalesced into a single memory transaction of 32, 64, or 128 bytes. Coalescing is achieved even if the warp is divergent and some threads of the half-warp do not actually access memory. Global memory access by all threads of a half-warp is coalesced into one or two memory transactions if all threads access the words in sequence: The kth thread in the half-warp must access the kth word. Otherwise, a separate memory transaction is issued for each thread and throughput is significantly reduced.

CUDA is a heterogeneous programming model as it assumes that the CUDA threads execute on a physically separate device that operates as a coprocessor to the host running the C program and that both the host and the device maintain their own separate memory spaces in DRAM, referred to as host memory and device memory. A typical CUDA procedure starts by allocating space on the device's memory, then transferring the required data for computation to the device, then launching the device kernel, and finally transferring the result from the device to the host. Processing flow on CUDA is shown in Fig. 3.27, where it can be summarized as below:

1. Copy data from main memory to GPU memory.
2. CPU initiates the GPU compute kernel.
3. GPU's CUDA cores execute the kernel in parallel.
4. Copy the resulting data from GPU memory to main memory.

Fig. 3.27 Processing flow on CUDA

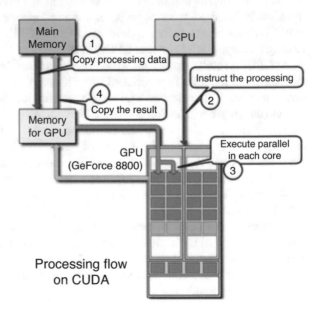

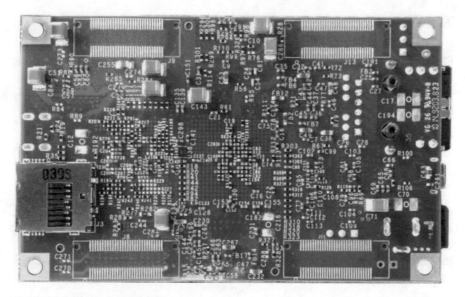

Fig. 3.28   Parallella board [24]

3.5.4   GPU Hardware

3.5.4.1   The Parallella Board

It supports parallel programming frameworks like MPI and OpenMP. The board kit
is shown in Fig. 3.28. The Parallella computer is a high-performance, credit card-
sized computer based on the Epiphany multicore chips from Adapteva. The Parallella
can be used as a standalone computer, as an embedded device, or as a component in
a scaled out parallel server cluster. The Parallella includes a low power dual core
ARM A9 processor and runs several of the popular Linux distributions, including
Ubuntu. The unique Epiphany coprocessor chips consists of a scalable array of simple RISC processors programmable in bare metal C/C++ or in a parallel programming frameworks like OpenCL [20], MPI, and OpenMP. The mesh of independent
cores are connected together with a fast on chip network within a distributed shared
memory architecture. The Parallella is an ideal candidate for anyone interested in
the field of parallel computing.

3.6   Parallel Computing: Overheads

Parallelism overheads include:

- Cost of starting a thread or process.
- Cost of communicating shared data.
- Cost of synchronizing.
- Extra (redundant) computation.

3.7   Parallel Computing: Performance

Parallel computing can speed-up simulations and get the solution faster by reducing the runtime. The speedup depends on the amount of code that cannot be parallelized. According to Amdahl's law which measures the parallelism degree, the speedup can be found by the following equation:

$$S_p = \frac{\text{Sequential execution time}}{\text{Parallel execution time}} = \frac{T_s}{T_P} = \frac{1}{\left(s + \dfrac{1-s}{n}\right)} \tag{3.1}$$

where: n is the number of processors, s is the fraction of program (algorithm) that is serial and cannot be parallelized, and $(1-s)$ is fraction parallelizable (Fig. 3.29) .Gustafson's law assumes that parallelism increases as the size of the data set increase. So, it is more optimistic than Amdahl's law. The speedup can be found by the following equation:

$$\text{Speedup} = n + (1-n)s \tag{3.2}$$

There is an overhead (synchronization, dependencies, communicating shared data), so the speedup is not just a linear function of number of processors.

A theoretical upper limit on parallel speedup is provided by Amdahl's law which assumes that communications is without cost. Actually, communications will further degrade the performance as depicted in Fig. 3.30.

There are many factors that can limit the speedup factor such as when the problem size is too small to take best advantage of a parallel computer. Speedup is limited when there is too much sequential code. Speedup is limited when there is too much parallel overhead compared to the amount of computation. Moreover, speedup is limited when the processors have different workloads.

Fig. 3.29 Program consists of serial part and parallel part

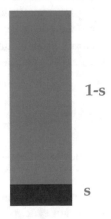

1-s

s

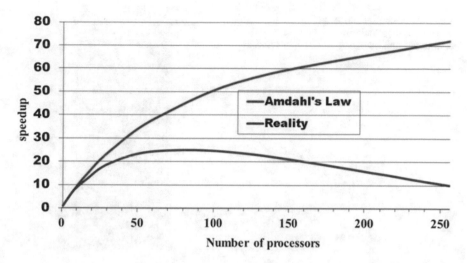

Fig. 3.30 Amdahl's law: practical and reality

Table 3.4 Types of granularity

Fine-grain parallelism (**GPUs**)	Coarse-grain parallelism (**clusters**)
• Low computation to communication ratio (Fig. 3.31) • Facilitates load balancing. • Implies high communication overhead and less opportunity for performance enhancement • Few sequential instructions per process • Decompose into small tasks (Fig. 3.32)	• High computation to communication ratio (Fig. 3.31) • Implies more **opportunity** for performance increase • Harder to load balance efficiently • Many sequential instructions per process • Decompose into large tasks (Fig. 3.32)

Granularity is considered as a qualitative measure of the ratio of computation to communication. Increasing granularity reduces the parallelism [21–23]. Machine granularity is given by:

$$\text{Machine granularity} = \frac{\text{Max number of computations which can be executed in 1s}}{\text{Max size message which can be transferred in 1s}} \quad (3.3)$$

There are two types of granularity that are summarized in Table 3.4

A performance comparison between three parallel programming models (MPI, Open MP, and CUDA) are presented. Open MP includes C/C++ based library routines compiler directives. Parallelism is supported by using shared memory model [25, 26].

The Open Message Passing Interface (Open MPI) supports the multithreading approach. It is used with applications that support concurrent access to memory [27].

Fig. 3.31 Fine granularity
versus coarse granularity:
communication overhead

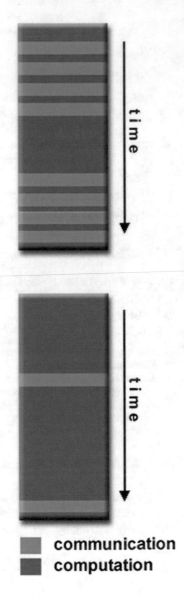

Compute Unified Device Architecture (CUDA) is developed by NVIDIA. CUDA runs on a graphical processing unit. GPUs are used for both graphics and non-graphic processing applications [28].

Cryptography algorithms are considered one of the most complex algorithms. The steps for constructing a parallel algorithm can be summarized as follows:

- Identify what pieces of work can be performed concurrently as not all applications can be parallelized.

coarse-grained   fine-grained

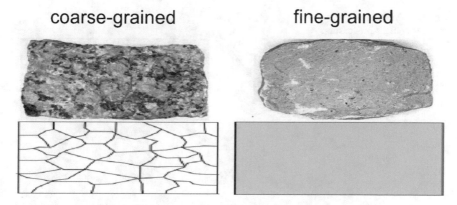

Fig. 3.32 Fine granularity versus coarse granularity: task decomposition

Fig. 3.33 Task partitioning diagram

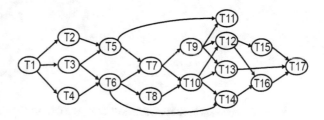

- Partition concurrent work onto independent processors—The mapping of work to processors can be done statically by the programmer or dynamically by the runtime. Task dependency diagram is used for partitioning (Fig. 3.33). Mapping example using 7 processors for 14 tasks is shown in Fig. 3.34.
    - **Start nodes: nodes with no incoming edges (T1).**
    - **Finish nodes: nodes with no outgoing edges (T17).**
    - **Dependency**: **T2 should be implemented after T1.**
    - **Critical path: the longest directed path between any pair of start and finish nodes.**
    - **Average degree of concurrency = total amount of work/ critical path length.**
- From the programmer point of view, there are three levels of parallelism: task-level parallelism, data-level parallelism, and Bit-level parallelism.
- Coordinate accesses to shared data: avoid conflicts.
- Ensure proper order of work using synchronization.

A comparison between the three languages is shown in Table 3.5. A comparison between the three languages in terms of speedup factor can be shown in Fig. 3.35, where CUDA shows the best performance. The main aim of this paper is to analyze the performance of different parallel programming languages. Cryptography example

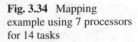

Fig. 3.34 Mapping example using 7 processors for 14 tasks

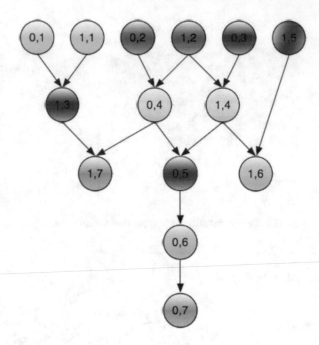

Table 3.5 Comparison between OPENMP, MPI, and CUDA

	OpenMP	MPI	CUDA
Used with	CPU	CPU	GPU
Based on	Compilation directives	Function calls	Function calls
Architecture	MIMD (shared memory)	MIMD (distributed/shared memory)	SIMD
Communication model	Shared address	Message passing	Shared address
Synchronization	Implicit	Implicit/explicit	Implicit
Implementation	Compiler	Library	Compiler

has been used. The parallel programming models significantly improve the performance in terms of run-time. The results show that CUDA has the best performance. GPU provides hundreds of cores vs. 2, 4, 8, 32 cores on CPU which can increase speed by up to 100x. Time elapsed and energy consumption comparison is shown in Fig. 3.36.

There are some performance limitations:

- *Wrong problem size*

  - Speedup is limited when the problem size is too small to take best advantage of a parallel computer.
  - In addition, speedup is limited when the problem size is fixed.

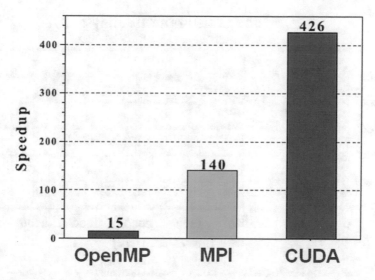

Fig. 3.35 Speedup factor comparison between the three languages. Cryptography example is used during this study

- That is, when the problem size does not grow as you compute with more processors.
- *Too much sequential code*
  - Speedup is limited when there is too much sequential code.
  - This is shown by Amdahl's law.
- *Too much parallel overhead*
  - Speedup is limited when there is too much parallel overhead compared to the amount of computation.
  - These are the additional CPU cycles accumulated in creating parallel regions, creating threads, synchronizing threads, spin/blocking threads, and ending parallel regions.
- *Load imbalance*
  - Speedup is limited when the processors have different workloads.
  - The processors that finish early will be idle while they are waiting for the other processors to catch up.

Efficiency gives fraction of time that processors are being used on computation as shown in the below equation:

$$\text{Efficiency} = \frac{\text{speedup}}{\#\,\text{processors}} \tag{3.4}$$

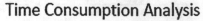

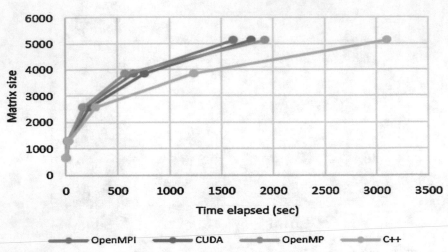

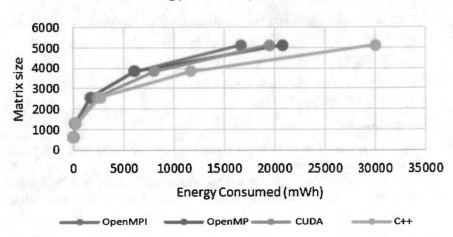

Fig. 3.36 Time elapsed and energy consumption comparison

3.8 New Trends in Parallel Computing

3.8.1 3D Processors

3D stacked multicore processor is one of the important applications of TSV-based 3D integration technology. Compared with standard 2D multicore processors, 3D integration allows increasing number of cores while reducing footprint area. For performance evaluation of 3D microprocessor architecture. MIPS (Million Instructions per Second) are used as an estimator. MIPS are calculated using the equation:

$$MIPS = IPC \times F_{max} \qquad (3.5)$$

IPC (Instructions Per Cycle) is calculated from cycle accurate processor simulation, and F_{max} is (maximum safe clock frequency).

Multicore processors stacking in the 3D design, where a 2D chip is divided into a number of different blocks and each one is placed on a separate layer of silicon where each layer is stacked on top of each other [29].

3.8.2 Network on Chip

NoCs have been introduced as a shared communication medium that is highly scalable and can offer enough **bandwidth** to replace many traditional bus-based and point-to-point links. NoC is a communication infrastructure for complex SoC systems with many IPs like a multiprocessor system. Instead of having a shared bus, a message passing approach is used. In analogy to a computer network, each IP acts like a processor, sending and receiving packets (flits) to and from the network. The basic element of a NoC is the switch, which connected to the IP with a network interface (NI). NoC architecture is shown in Fig. 3.37.

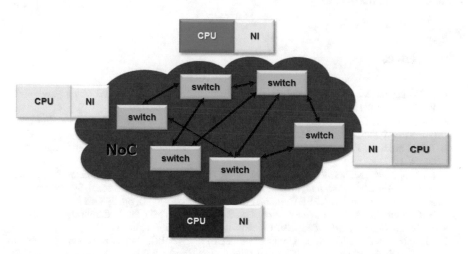

Fig. 3.37 NoC

3.8.3 FCUDA

It is a synthesis tool that enables efficient compilation of CUDA kernels onto FPGAs. It provides better performance. GPUs and FPGAs are becoming very popular in PC-based heterogeneous systems for speeding up compute intensive kernels of scientific, imaging, and simulation applications. GPUs can execute hundreds of concurrent threads, while FPGAs provide customized concurrency for highly parallel kernels. However, exploiting the parallelism available in these applications is currently not a push-button task. Often the programmer has to expose the application's fine and coarse grained parallelism by using special APIs. CUDA is such a parallel computing API that is driven by the GPU industry and is gaining significant popularity [30].

3.9 Conclusions

One of the ultimate goals of improving computing is to increased performance without increasing clock frequencies and to overcome the power limitations of the dark-silicon era. In this chapter, parallel computing is introduced. A performance comparison of MPI, Open MP, and CUDA parallel programming languages are presented. The performance is analyzed using cryptography as a case study. The results show that CUDA has the best performance in terms of runtime. The recent advances in parallel computing capabilities have enabled the deployment of data-driven DL approaches to complement the conventional model-based approaches.

References

1. B. Wilkinson, M. Allen, *Parallel Programming Techniques & Applications Using Networked Workstations & Parallel Computers*, 2nd edn. (Pearson, Toronto, 2004)
2. A. Grama, A. Gupta, G. Karypis, V. Kumar, *Introduction to Parallel Computing*, 2nd edn. (Addison Wesley, Reading, MA, 2003)
3. T.G. Lewis, H. El-Rewini, *Introduction to Parallel Computing* (Prentice Hall, Englewood Cliffs, 1992)
4. J. Zhang, T. Ke, M. Sun, *The Parallel Computing Based on Cluster Computer in the Processing of Mass Aerial Digital Images*. In: F. Yu and Q. Lou, Eds. International Symposium on Information Processing (IEEE Computer Society, Moscow, 2008), pp. 398–393
5. N. Carriero, D. Gelernter, *How to Write Parallel Programs* (MIT Press, Cambridge, 1990)
6. M. Flynn, *Encyclopedia of Parallel Computing* (Springer, Berlin, 2011)
7. Riscv.org, KISS PULPino. Updates on PULPino (2019). https://riscv.org/wp-content/uploads/2016/12/Tue0915-RISC-V-PULPino-update-Zaruba-ETH-Zurich.pdf
8. Actel.com, LEON3/LEON3-FT CompanionCore Data Sheet (2019). http://www.actel.com/ipdocs/leon3_ds.pdf. (Accessed Sep 2019)
9. En.wikipedia.org, MicroBlaze (2019). https://en.wikipedia.org/wiki/MicroBlaze

10. Intel.com, Nios II Processors for FPGAs (2019). https://www.intel.com/content/www/us/en/products/programmable/processor/nios-ii.html
11. Digilentinc.com, The usefulness of Soft CPU Cores in FPGA Devices(2019). https://forum.digilentinc.com/topic/4735-the-usefullness-of-soft-cpu-cores-in-fpga-devices
12. P. Liu, C.H. Yang, Locality-preserving dynamic load balancing for data parallel applications on distributed memory multiprocessors. J. Inf. Sci. Eng. **18**(6), 1037–1048 (2002)
13. P. Caheny, M. Casas, M. Moreto, et al., *Reducing Cache Coherence Traffic with Hierarchical Directory Cache and NUMA-Aware Runtime Scheduling*. In: 2016 International Conference on Parallel Architecture and Compilation Techniques (PACT) (2016), pp. 275–286
14. M. Dashti, A. Fedorova, J. Funston, et al., Traffic management: A holistic approach to memory placement on NUMA Systems. In: Proceedings of the Eighteenth International Conference on Architectural Support for Programming Languages and Operating Systems. ASPLOS'13 (ACM, New York, 2013), pp. 381–394
15. http://openmp.org/wp/
16. www.mpi-forum.org/docs/docs.html
17. K. Salah, M. AbdelSalam, *A Comparative Analysis Between FPGA and GPU for Solving Large Numbers of Linear Equations*. Microelectronics (ICM), 2017 29th International Conference on IEEE (2017)
18. http://docs.nvidia.com/cuda
19. D. B. Kirk, W. W. Hwu, *Programming Massively Parallel Processors—A Hands-On Approach* (Morgan Kaufmann)
20. Intel, Intel FPGA SDK for OpenCL—Programming Guide (2017). UGOCL002. http://bit.ly/2blgksq
21. J. Enos, et al., *Quantifying the Impact of GPUs on Performance and Energy Efficiency in HPC Clusters*, In: Green Computing Conference, 2010 International (IEEE, 2010)
22. M. Rashid, L. Ardito, M. Torchiano, *Energy Consumption Analysis of Algorithms Implementations*, In: Empirical Software Engineering and Measurement (ESEM), 2015 ACM/IEEE International Symposium on (IEEE, 2015)
23. S. Roy, A. Rudra, A. Verma, *An Energy Complexity Model for Algorithms*, In: Proceedings of the 4th conference on Innovations in Theoretical Computer Science (ACM, 2013)
24. https://www.parallella.org/board/
25. T. Chen, Y.-K. Chen, Challenges and opportunities of obtaining performance from multi-core cpus and many-core gpus, In *Acoustics, Speech and Signal Processing*, 2009. IEEE International Conference on, April 2009 (ICASSP, 2009), pp. 613–616
26. L. Dagum, R. Menon, OpenMP: An industry standard API for shared-memory programming. Computational Science & Engineering, IEEE **5**(1), 46–55 (1998)
27. E. Gabriel et al., Open MPI: Goals, concept, and design of a next generation MPI implementation, in *Recent Advances in Parallel Virtual Machine and Message Passing Interface*, (Springer, Berlin, 2004), pp. 97–104
28. D. Kirk, *NVIDIA CUDA Software and GPU Parallel Computing Architecture*, vol. 7 (ISMM, 2007)
29. K. Salah, A. El Rouby, H. Ragai, Y. Ismail, *3D/TSV Enabling Technologies for SOC/NOC: Modeling and Design Challenges*. In 2010 International Conference on Microelectronics (pp. 268–271) (IEEE, 2010)
30. J. Stratton et al., *MCUDA: An Efficient Implementation of CUDA Kernels for Multi-Core CPUs*. Int. Workshop on Languages and Compilers for Parallel Computing (2008)

Chapter 4
Deep Learning and Cognitive Computing: Pillars and Ladders

4.1 Introduction

Deep learning has become one of the most important computationally intensive applications for a wide variety of fields. Deep learning is a subfield of machine learning that deals with algorithms inspired by the structure and function of the brain. Machine learning is a subset of artificial intelligence as shown in Fig. 4.1. Deep neural networks (DNNs) are a family of neuromorphic computing architectures that have recently made substantial advances in difficult machine learning problems such as image or object recognition, speech recognition and machine language translation. Now machine learning is everywhere. Facebook as an example, machine learning in search, face tagging, advertisement, and news feed.

4.1.1 Artificial Intelligence

The definition of artificial intelligence (AI) is very broad. Theoretically, it referrers to the technology that machine can think and act spontaneously without specifically programming for it. In such way, robots can work for human and be able to undertake different types of jobs. AI is the one of the revolutionary technologies which will bring great changes to the economics. AI can be a pile of if-else statements or a complex statistical model. It is all about decision-making and reasoning. AI is the ability to sense, reason, engage, and learn. **AI = Machine learning + voice recognition + natural language processing + computer vision + knowledge capture + planning and optimization + self-correction.**

AI, sometimes called machine intelligence, is intelligence demonstrated by machines, in contrast to the natural intelligence displayed by humans and other animals. In computer science AI research is defined as the study of intelligent agent which are any device that perceives its environment and takes actions that maximize

© Springer Nature Switzerland AG 2020
K. S. Mohamed, *Neuromorphic Computing and Beyond*,
https://doi.org/10.1007/978-3-030-37224-8_4

Fig. 4.1 Deep learning is a subfield of machine learning that deals with algorithms inspired by the structure and function of the brain. **Venn diagram** of the relation between deep learning, machine learning, and AI

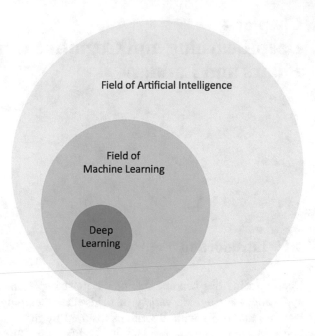

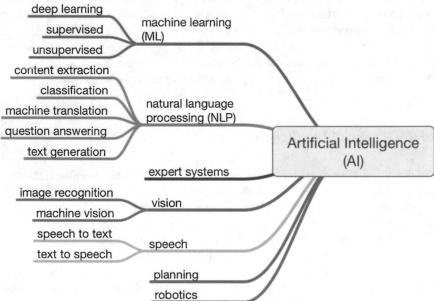

Fig. 4.2 AI

its chance of successfully achieving its goals (Fig. 4.2). AI is applied when a machine mimics "cognitive" functions that humans associate with other human minds, such as "learning" and "problem solving" [1]. There are different types of AI:

Table 4.1 History of machine learning

Year	Achievement
1958	Frank Rosenblatt designed the first artificial neural network called perceptron.
1982	John Hopfield suggested creating a network which had bidirectional lines, similar to how neurons actually work.
2012	Google Brain: This was a deep neural network created by Jeff Dean of Google, which focused on pattern detection in images and videos. It was able to use Google's resources, which made it incomparable to much smaller neural networks. It was later used to detect objects in YouTube videos.

- **Weak AI** (narrow AI): It is an AI system that is designed and trained for a particular task. It is trained and developed to perform this task and cannot do unfamiliar tasks to it which has no previous training experience with it.
- **Strong AI** (artificial general intelligence): It is AI system which has its own conscious; it has the ability to find solution without human intervention when it is exposed to unfamiliar task. It can make a decision, self-awareness, and learn from itself. It is not practical yet.

4.1.2 Machine Learning

History of machine learning has been summarized in Table 4.1. The core technology in AI is machine learning (ML). Intrinsically, ML is algorithms that can be applied to dataset to perform the tasks of *clustering, classification, and regression*. Clustering is used to sort out the similarities of the data without knowing the structure of the data; classification is used to sort the new data based on the known data structure; regression is to find the correlation of different variables and build the prediction model to forecast the data trend. It is very similar as the tasks in Data Mining, but there is a slight difference. ML is more focused on the development of the algorithms so that the computer can automatically improve along with experience. However, Data Mining emphasizes on the application of the algorithms and uses it as tools to extract valuable patterns in the datasets. ML can be divided into three main branches: supervised, unsupervised and reinforcement machine learning. Comparison between the different types of machine learning in terms of feedback is shown in Table 4.2. There are classical methods for machine learning and adaptive methods. Classical system receives some input values, processes them, and produces output results. Adaptive system has a feedback to train and tune such as neural networks machine learning. Some of classical machine learning algorithms are:

- **Decision Trees**: A decision tree is a decision support tool that uses a tree-like graph or model of decisions and their possible consequences, including chance-event outcomes, resource costs, and utility. It allows you to approach the problem in a structured and systematic way to arrive at a logical conclusion.

Table 4.2 Comparison between the different categories of machine learning in terms of feedback

Supervised	Give feedback on right on wrong, and adapt to the feedback. • **Classification**: predict a category. • **Regression**: predict a number.
Unsupervised	Give no feedback, the machine should learn the structure of the data to solve the task. • **Clustering**: divide by similarity. • **Association**: identify sequences. • **Dimension reduction**: find hidden dependencies.
Reinforcement	Isn't giving feedback, only receive feedback if it achieve the goal. • **Recommendation systems**.

- **Naive Bayes Classification**: It is simple probabilistic classifiers based on applying Bayes' theorem with strong (naive) independence assumptions between the features. The featured image is the equation—$P(A|B)$ is posterior probability, $P(B|A)$ is likelihood, $P(A)$ is class prior probability, and $P(B)$ is predictor prior probability.
- **Least Squares Regression**: Regression as the task of fitting a straight line through a set of points. A line is drawn, and then for each of the data points, measure the vertical distance between the point and the line, and add these up; the fitted line would be the one where this sum of distances is as small as possible.
- **Logistic Regression**: Logistic regression is a powerful statistical way of modeling a binomial outcome with one or more explanatory variables. It measures the relationship between the categorical dependent variable and one or more independent variables by estimating probabilities using a logistic function, which is the cumulative logistic distribution.
- **Support Vector Machines**: SVM is binary classification algorithm. Given a set of points of two types in N dimensional place, SVM generates a $(N-1)$ dimensional hyperplane to separate those points into two groups. Say you have some points of two types in a paper which are linearly separable. SVM will find a straight line which separates those points into two types and situated as far as possible from all those points.
- **K Nearest Neighbors:** is a simple algorithm that stores all available cases and classifies new cases based on a similarity measure (e.g., distance functions).

Rule-based systems and machine learning systems are not the same. A rule-based system requires a human to find patterns in the data and define a set of rules that can be applied by the algorithm. The rules are typically a series of if-then-else statements that are executed in a specific sequence. A machine learning system, on the other hand, discovers its own patterns and can continue to learn with each new prediction on unseen data.

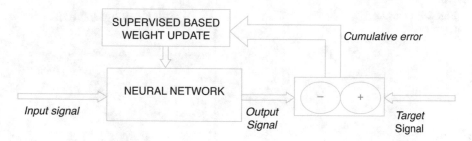

Fig. 4.3 Supervised machine learning

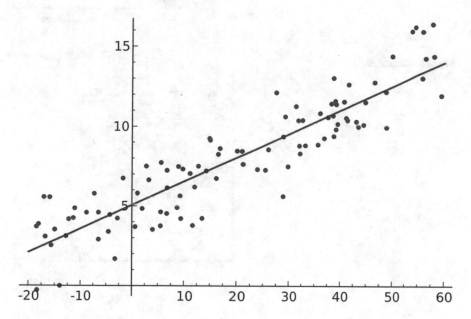

Fig. 4.4 Linear Regressions

4.1.2.1   Supervised Machine Learning

If for a certain input the corresponding output is known, the network is to learn the mapping from inputs to outputs (Fig. 4.3). Supervised ML is error based. Classification and regression are examples for the problems which can be solved by supervised machine learning. Figure 4.4 shows a linear **regressions** example. Regression can be linear or nonlinear (Fig. 4.5). Regression analysis refers to methods of modeling and analyzing several variables where the focus is on the relationship between a dependent variable and one or more independent ("causing") variables. Figure 4.6 shows a **classification** example. Classification can be for single object or multi-objects

Fig. 4.5 Logistic
regressions

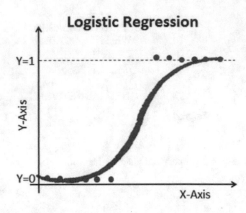

Fig. 4.6 Classification

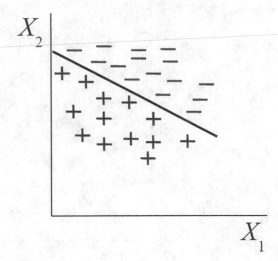

4.1.2.2 Unsupervised Machine Learning

In unsupervised learning, there is no a priori output and no information is given from an outside observer to advise or correct (Fig. 4.7). Rather, a model is built and adjusted so that it fits the observed data. Unsupervised ML is memory-based algorithm. *Clustering is* an example for the problems which can be solved by unsupervised machine learning. Figure 4.8 shows **a clustering** example. Moreover, unsupervised can be used in **association** and **dimension reduction** problems. Association means finding items that tend to co-occur in the data and specifies the rules that govern their co-occurrence. One of the applications of dimension reduction is compression.

Back-propagation is considered the standard method in artificial neural networks to calculate the error contribution of each neuron after a batch of data is processed. However, there are some major problems using back-propagation. Firstly, it requires labeled training data, while almost all data is unlabeled.

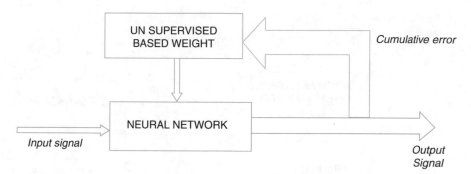

Fig. 4.7 Unsupervised machine learning

Fig. 4.8 Clustering

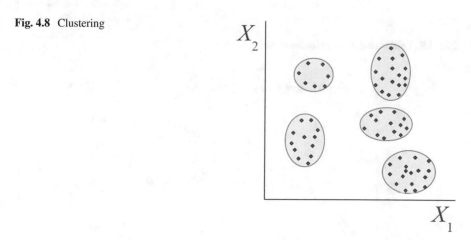

Secondly, the learning time does not scale well, which means it is very slow in networks with multiple hidden layers. Thirdly, it can get stuck in poor local optima, so for deep nets they are far from optimal.

To overcome the limitations of back-propagation, researchers have considered using unsupervised learning approaches. This helps keep the efficiency and simplicity of using a gradient method for adjusting the weights, but also use it for modeling the structure of the sensory input. In particular, they adjust the weights to maximize the probability that a generative model would have generated the input.

4.1.2.3   Reinforcement Machine Learning

Reinforcement machine learning is used to take actions and in recommendation systems. If the teacher only tells a student whether her answer is correct or not, but leaves the task of determining why the answer is correct or false to the student, we have an instance of reinforcement learning. Reinforcement learning can be based on value or policy (Fig. 4.9). One of the applications is robot navigation systems.

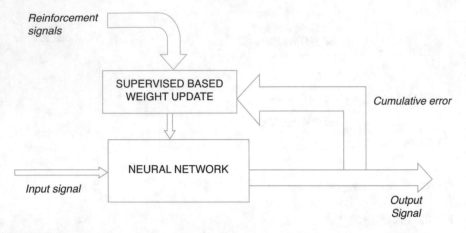

Fig. 4.9 Reinforcement machine learning

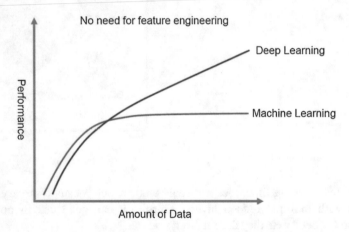

Fig. 4.10 Amount of data versus performance for machine learning and deep learning [3]

4.1.3   *Neural Network and Deep Learning*

Artificial neural network is an important machine learning algorithm. Neuron is the basic unit in the human's brain. It is used to process the information and transmit it to the next neuron. The connection between different neurons forms the human brain network structure. Artificial neural network (ANN) is the analogy to human's brain, with which machine can perform the tasks of classification and pattern recognition [2]. Deep learning indicates the ANN structure with at least one hidden layer. The more the hidden layers, the deeper the learning is (Fig. 4.10). Complicated task often requires more than one hidden layer as well as more neurons in each layer. Different generations of NNs have been summarized in Table 4.3. Deep learning uses especially powerful neural networks. Deep networks refer to networks with

Table 4.3 Different generations of NNs

Generation	Type	Description
1st	NN	One hidden layer
2nd	2D/3D CNN	Many hidden layers
3rd	Spiking NN	Efficient in terms of area and energy

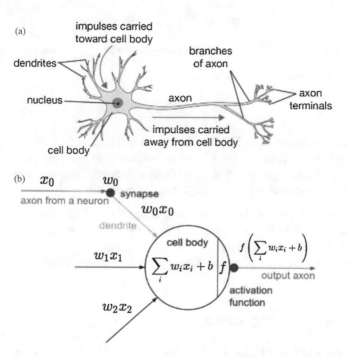

Fig. 4.11 Analogy between a biological neuron (perceptron) (**a**) and its mathematical model (**b**). Your brain is jam packed full of neurons like these in figure and neurons are cells in the brain. Neuron consists of: Dendrite: it is the input wires of neuron which used to interface with other neurons in case of communication. Cell body: its body used to do some computations. Axon: output wires form neurons, May are used to communicate with the input wires of another neuron or send messages in form of pulses, it will make group of them to have a network

more than three layers. The others are called **shallow** networks [4–7]. Deep learning is a subfield of machine learning that deals with algorithms inspired by the structure and function of the brain. **Venn diagram** of the relation between deep learning, machine learning, and AI is shown in Fig. 4.1. Analogy between a biological neuron and its mathematical model is shown in Fig. 4.11. NNs development is shown in Fig. 4.12 [8, 9].

A single-layer artificial neural network, also called a single-layer, has a single layer of nodes, as its name suggests. Each node in the single layer connects directly to an input variable and contributes to an output variable. A single-layer network can be extended to a multiple-layer network, referred to as a Multilayer Perceptron.

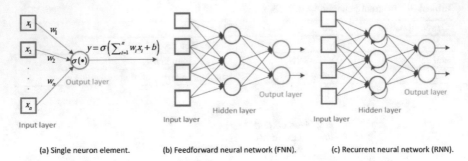

(a) Single neuron element. (b) Feedforward neural network (FNN). (c) Recurrent neural network (RNN).

Fig. 4.12 NNs: development

A Multilayer Perceptron, or **MLP** for sort, is an artificial neural network with more than a single layer. It has an input layer that connects to the input variables, one or more hidden layers, and an output layer that produces the output variables. We can summarize the types of layers in an MLP as follows [10]:

- Input Layer: Input variables, sometimes called the visible layer.
- Hidden Layers: Layers of nodes between the input and output layers. There may be one or more of these layers.
- Output Layer: A layer of nodes that produce the output variables.

4.2 Deep Learning: Basics

4.2.1 DL: What? Deep vs. Shallow

The Deep Learning (DL) is a set of machine learning algorithms that try to simulate the high-level abstraction of data, using architecture, consisting of a plurality of non-linear transformations. The term "depth" in this case refers to the depth of the graph model of computation—the maximum length between the input and output nodes of a particular architecture. The term "deep learning" focuses on the complexity of the training of internal (deep) layers of the multilayer networks that are difficult to train using classical teaching methods such as the method of back-propagation.

According to the number of hidden layers, the neural network can be classified as Deep Neural Network (**DNN**) which usually has more than two layers and shallow neural network. In terms of the direction of data flow, there is Recurrent Neural Network [11].

In terms of the direction of data flow, there are Recurrent Neural Network (**RNN**) where there is a data flow between adjacent cells in the same layer and general Neural Network where there is not such connection. Judged from the operations

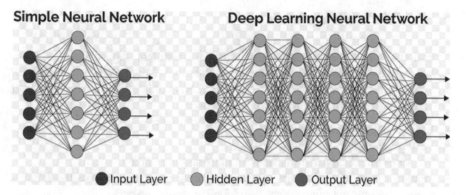

Fig. 4.13 The main difference between deep neural networks and conventional neural networks is the number of hidden layers as deep neural network contains larger number of hidden layers

Fig. 4.14 Human visual cortex

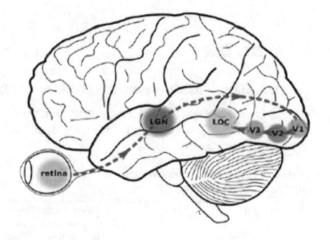

executed by the neural network, there is a Convolutional Neural Network (**CNN**) which mainly performs its function through convolutional operation. The main difference between deep neural networks and conventional neural networks is the number of hidden layers as deep neural network contains larger number of hidden layers or more than one hidden layer as shown in Fig. 4.13. Sometimes, one hidden layer is hard to train some types of nonlinear data. There is a type of CNN called binary-weight CNN where it executes only addition operations, in which weights are constrained to binary values (+1 and −1). Human visual cortex is shown in Fig. 4.14.

4.2.2   DL: Why? Applications

Neural networks have different architectures for different applications. Table 4.4 shows different NNs architectures versus applications. Mainly, we have three main streams for deep learning (CNN, RNN, and GAN) and they will be discussed later. Different applications are summarized as follows:

- Process huge amount of data.
- Perform complex algorithms.
- To achieve the best performance with large amount of data.
- Feature extraction: take large volumes of data as input, analyze the input to extract features out of an object, and identify similar objects.
- Cancer detection.
- Robot navigation.
- Autonomous driving: distinguish between different types of objects, people, and road signs.

4.2.3   DL: How?

Deep learning consists of two modes (Figs. 4.16 and 4.17):

- Training mode: learn to predict the output from the input.
- Inference mode: predict the output from the input. This input is never seen before during training mode.

Table 4.4   Different NNs versus Applications

Classification	Type	Application
Feed-forward	CNN	Object recognition Image segmentation Handwriting digit recognition (Fig. 4.15)
Recurrent	RNN	Speech processing Text processing
Generative	GAN	Photo synthesis Video synthesis

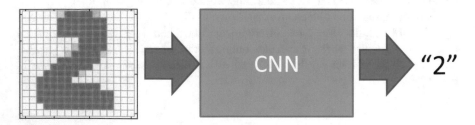

Fig. 4.15 Handwriting digit recognition

Fig. 4.16 Deep learning flow (abstract)

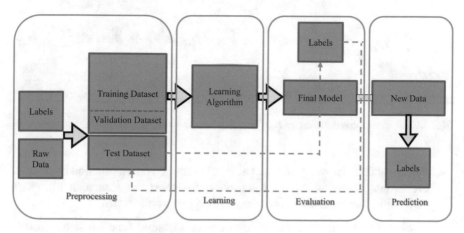

Fig. 4.17 Deep learning flow (in details)

The governing equation is as follows (**MAC** operation):

$$u(x) = \sum_{i=1}^{n} w_i x_i + b \qquad (4.1)$$

$u(x)$ is the predicted output
x_i is the input
w_i is the weights
b is the bias

Equation (4.1) is a linear function. DL can also be described as nonlinear function for nonlinear data such as:

$$y = \text{activation}(u) \qquad (4.2)$$

Figure 4.18 shows how DL network works in the following steps:

1. Random weights and bias are initialized. The bias is also a weight. The bias try to catch non-observable factors.
2. **Training data** such as images are inserted as input to the network. Use the training data to find the network parameters.
3. Calculate the weighted sum of the inputs.

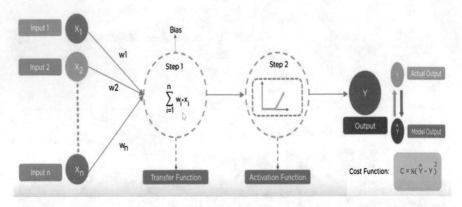

Fig. 4.18 A motivational example: how neural networks work. Learning is about parameters tuning

4. Pass the calculated weighted sum as input to the activation function to generate the output. Different types of activation functions are shown in Fig. 4.19. Activation functions allow the system to capture nonlinearity.

- Sigmoid or Logistic Activation Function: Sigmoid function is between (0 and 1). Therefore, it is especially used for models where we have to predict the probability as an output. Since probability of anything exists only between the range of 0 and 1, sigmoid is the right choice. The function is differentiable (we can find its slope at any two points). The function is monotonic but function's derivative is not.
- Softmax Function is a more generalized logistic activation function which is used for multiclass classification.
- Tanh or hyperbolic tangent Activation Function: It is like logistic sigmoid but the range of the tanh function is from (−1 to 1). The advantage is that the negative inputs will be mapped strongly negative and the zero inputs will be mapped near zero in the tanh graph. The function is differentiable. The function is monotonic while its derivative is not monotonic. The tanh function is mainly used in classification between two classes.
- ReLU (Rectified Linear Unit) Activation Function: The ReLU is the most used activation function in the world right now. Since, it is used in almost all the convolutional neural networks or deep learning. The ReLU is half rectified (from bottom). f(z) is zero when z is less than zero and f(z) is equal to z when z is above or equal to zero (Range from 0 to infinity). The function and its derivative both are monotonic.
- Leaky ReLU:
- The leak helps to increase the range of the ReLU function. ReLU functions and their derivatives are both monotonic.

5. Calculate the error (cost function), if it is acceptable then the network is trained, else change the values of the weights using **back-propagation** method and repeat steps 1–5. An example for error function versus iterations is shown in Fig. 4.20.

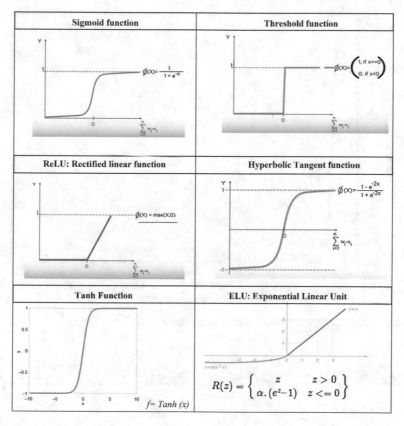

Fig. 4.19 Types of activation functions. Any nonlinear function can be approximated to linear function by piecewise linear method

$$\text{Cost function} = \sum_{i=1}^{n} \left(\text{Prediction}_i - \text{Target}_i \right)^2 \qquad (4.3)$$

6. Our target is to reach the minimum value for the cost function using **gradient descent update rule** which is an optimization technique used to find the minimum of a function(It is an iterative optimization algorithm used to find values of Weights and Bias that minimize the Cost Function). A typical way to learn a deep learning model is to back-propagate the loss gradients through the model and update its weights by a gradient descent optimizer. The loss is produced through feedforward propagation and then it is back-propagated to produce each layer's weight gradients. In the whole training process consisting of the forward and backward phase, there are three types of computations, which are feedforward propagation, back-propagation, and weight gradient computation. Figure 4.21 shows Forward propagation versus back-propagation. Figure 4.22 shows the cost function of the weights, where the gradient descent optimization

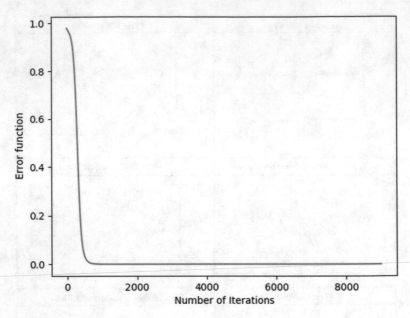

Fig. 4.20 Error cost function versus number of iterations

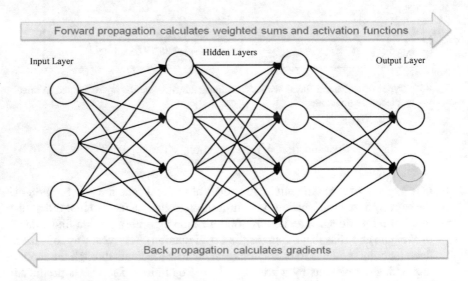

Fig. 4.21 Forward propagation versus back-propagation in a multiple hidden layers example. A typical way to learn a deep learning model is to back-propagate the loss gradients through the model and update its weights by a gradient descent optimizer. The loss is produced through feed-forward propagation and then it is back-propagated to produce each layer's weight gradients

Fig. 4.22 Gradient: towards the global cost minimum (w) is the cost function

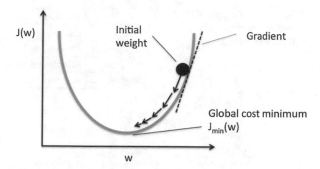

algorithm can be used to find the weights that minimize the cost function to classify the input samples [12–14]. Updating the weight is governed by the below equation:

$$W_{new} = W_{old} - \alpha \frac{\partial J}{\partial W_{old}} \tag{4.4}$$

where α is the learning rate which determines how big or small the update will be.
7. We can use pruning, quantization, and compression.

- Pruning/approximation: setting near-zero weights to zero.
- **Quantization**: replacing 32-bit floats with 8-bit integers.
- Compression: storing sparse weight matrix in compressed format.

8. Finally, the network enters the **testing phase** in which test data (that it did not process before) is entered as inputs and then the network outputs its prediction/inference. It is a pretrained model.
9. If the prediction is wrong we need to repeat the training phase again with more training data.
10. Table 4.5 shows a comparison between training phase and inference phase.

4.2.4 DL: Frameworks and Tools

A lot of Machine Learning/Deep Learning platforms are available. A ML/DL platform is a library to help you develop and train ML/DL models and provide a better environment for that matter. Deep learning frameworks are sets of software libraries that implement the common training and inference operations. Examples of these include **Caffe**/Caffe-2 (Facebook) [15], **TensorFlow** (Google) [16], Torch [17], PyTorch [18], cognitive toolkit (Microsoft) [19], MxNet (Amazon) [20], and **KERAS** (Python based) [21]. All of these are available as open-source software. Such frameworks are supported across both GPUs and FPGAs. Choosing DL framework depends on easy of programming and running speed (Fig. 4.23). Machine learning as a service (MLaaS) is a new and promising trend. Table 4.6 summarizes the most famous tools in DNNs.

Table 4.5 Comparison between training phase and inference phase

	Training	Inference
Dataset	Large	One sample at a time
Propagation	Forward and backward	Forward only.
Run-time	Minutes up to weeks	m-seconds
Weights	Are computed	Are known

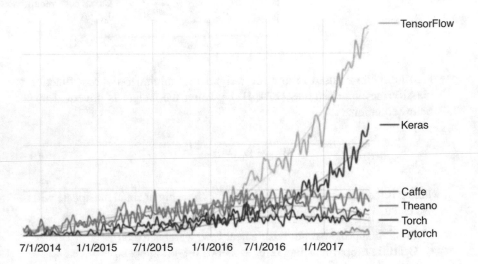

Fig. 4.23 Deep learning framework progress

Table 4.6 Different DNN tools

Tool	Description
OpenCV [22]	Computer vision library in C++
TensorFlow	Open-source software library for machine learning developed by Google
Keras	Keras is a deep learning framework for Python that provides a convenient way to define and train almost any kind of deep learning model

4.2.4.1   TensorFlow

TensorFlow is a free and open-source software library for dataflow and differentiable programming across a range of tasks. It is a symbolic math library and is also used for machine learning applications such as neural networks. The name TensorFlow derives from the operations that such neural networks perform on multidimensional data arrays, which are referred to as tensors. Tensors are a generalization of vectors (matrices) to an arbitrary number of dimensions [23, 24]

TensorFlow is a framework developed by Google that uses a static graph, which means building the graph once, then executing it many times. TensorFlow is a very-low-level numerical library which led to a number of different libraries that aim to provide a high-level abstraction layers such as Keras, Sonnet, TFLearn, and others. In TensorFlow, you can define Placeholders which are input nodes in the computational graph, such as the input data. Variables are values that exist within the computational graph that may be updated, such as weights and biases. The same graph can then be executed many times in a session. Static graphs have many advantages as mentioned before. Unfortunately, it also has some drawbacks that Google attempted to address recently. It can be a bit difficult to debug the code as it is not executed imperatively, and it has a huge overhead on prototyping which is not ideal for research work. Google introduced Eager Execution, an imperative, define-by-run interface for TensorFlow. However, this feature is still in its early stages.

4.2.4.2   Keras

Keras is a deep learning framework for Python that provides a convenient way to define and train almost any kind of deep learning model. Keras was initially developed for researchers, with the aim of enabling fast experimentation. It allows the same code to run seamlessly on CPU or GPU and it has a user-friendly API that makes it easy to quickly prototype deep learning models. Keras has built-in support for convolutional networks (for computer vision), recurrent networks (for sequence processing), and any combination of both. Python is a dynamic language. Dynamic languages can be interpreted directly, which means that the actual text of the program (the source code) is used while the program is running. In contrast, a static language is executed in two phases: first the program is translated from source code to binary code, and then the binary code is interpreted. Some key features of Keras are as follows:

- It allows the same code to run seamlessly on CPU or GPU.
- It has a user-friendly API that makes it easy to quickly prototype deep learning models.
- It has built-in support for convolutional networks (for computer vision), recurrent networks (for sequence processing), and any combination of both.
- It supports arbitrary network architectures: multi-input or multi-output models, layer sharing, model sharing, and so on. This means Keras is appropriate for building essentially any deep learning model, from a generative adversarial network to a neural Turing machine.
- Keras is distributed under the permissive MIT license, which means it can be freely used in commercial projects.
- It is compatible with any version of Python from 2.7 to 3.6.

4.2.4.3   PyTorch

PyTorch is a framework developed by Facebook that uses dynamic graphs which means building a new computational graph on each forward pass. It is quite similar to Torch, and shares some of its backend. PyTorch is deeply integrated with Python and follows an object-oriented paradigm. It also allows you to easily extend functionality by simply defining your own classes that extend PyTorch. For example, creating a custom neural network class extends nn.Module. The imperative nature of PyTorch makes it really easy to write clean code that is easy to debug, and utilize typical Python functionality such as conditionals and loops. PyTorch ships with three levels of abstractions to make things easier to use. A Tensor in PyTorch is an imperative nd-array, similar to numpy while having the ability to run on the GPU (Fig. 4.24). A Variable is a node in the computational graph, which is very similar to the Tensor, Variable, and Placeholder in TensorFlow. A Module is a neural network layer which can store weights or learned weights, which can be used to create your own neural network classes. While PyTorch is less mature compared to TensorFlow, it gained a lot of popularity within the research community due to its imperative nature and Pythonic API. There is a development community around the framework that includes various libraries such as visualization tools [25].

4.2.4.4 OpenCV

OpenCV (Open Source Computer Vision) is a library of programming functions mainly aimed at real-time computer vision, originally developed by Intel's research center. I is widely used for image processing. Officially launched in 1999, the OpenCV project was initially an Intel Research initiative to advance CPU-intensive applications, part of a series of projects including real-time ray tracing and 3D display walls. In the early days of OpenCV, the goals of the project were described as:

- Advanced vision research by providing open and optimized code for basic vision infrastructure.
- Disseminate vision knowledge by providing a common infrastructure that developers could build on, so that code would be more readily readable and transferable.

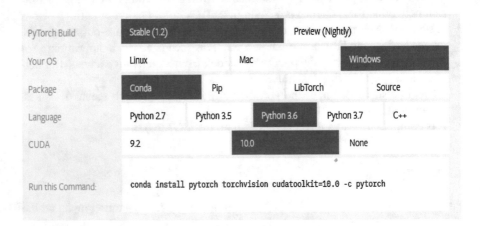

Fig. 4.24 PyTorch: The big picture

- Advance vision-based commercial applications by making portable, performance-optimized code available for free—with a license that did not require being open or freeing them.

4.2.4.5   Others

There are other frameworks such as Caffe, Theano, and Torch. Comparison between different deep learning frameworks are shown in Table 4.7.

4.2.5   DL: Hardware

Google built the Tensor Processing Unit (**TPU**), an ASIC designed from the ground up for machine learning that powers several of major products, including Translate, Photos, Search, Assistant, and Gmail (Fig. 4.25) [26]. NVidia built **Volta** which is a Tensor Core GPU architecture designed to bring AI to every industry [27]. Moreover, **Microsoft Brainwave** is a deep learning platform for real-time AI serving in the cloud [28]. MyHDL turns Python into a hardware description and verification language, providing hardware engineers with the power of the Python ecosystem. **MyHDL** is a free, open-source package for using Python as a hardware description and verification language. Python is a very-high-level language, and hardware designers can use its full power to model and simulate their designs. Moreover, MyHDL can convert a design to Verilog or VHDL. This provides a path into a traditional design flow. Subject to some limitations, MyHDL supports the automatic conversion of MyHDL code to Verilog or VHDL code. This feature provides a path from MyHDL into a standard Verilog or VHDL-based design environment to be convertible, the hardware description should satisfy certain restrictions, defined as the convertible subset. A convertible design can be converted to an equivalent model in Verilog or VHDL, using the function to Verilog or to VHDL from the MyHDL library. The converter attempts to issue clear error messages when it encounters a construct that cannot be converted [29]. In [30], high-performance DNN inference acceleration unit is proposed and it is called **ManyCore™** Structure.

4.3   Deep Learning: Different Models

4.3.1   Feedforward Neural Network

This is one of the simplest types of artificial neural networks. In a feedforward neural network, the data passes through the different input nodes till it reaches the output node. In other words, data moves in only one direction from the first tier onwards until it reaches the output node (connections between the nodes do not form a cycle).

Table 4.7 Comparison between different deep learning frameworks

Metric	Tensorflow	Keras	Pytorch
Definition	TensorFlow is an open-sourced library. It is one of the more famous libraries when it comes to dealing with Deep Neural Networks.	Keras is an open-source network library written in python. It is capable of running on top of TensorFlow. Developers can use Keras to quickly build neural networks without worrying about the mathematical aspects of tensor algebra, numerical techniques, and optimization methods.	Pytorch is a machine learning library based on the torch library, used applications such as computer vision and natural language processing
Speed	Faster and suitable for high performance	Slower	Faster same as Tensorflow
Level of API	Provides both high- and low-level API	High-level API	Provides lower-level API
Architecture	It is not very easy to use, so we can use Keras as framework that makes work easier	Has simple architecture and is "more readable"	Has complex architecture and the readability is less
Debugging	There is a special dedicated tool called Tensorflow debugger	Has a lot of computation junk in its abstractions and so it becomes difficult to debug	Allows an easy access to the code and it is easier to focus on the execution of each line
Dataset	Is used for high-performance models and large dataset that require fast execution	Is used for small datasets as it is the slowest	Same as TensorFlow
Popularity	The middle	Has the highest popularity	The lowest
Operating system	Android—ios—Mac os—windows—Linux and Raspberry Pi	Linus and OSX	Windows—Linux and OSX
Conclusions	Suitable for: • Large Dataset • High performance • Functionality	Suitable for: • Rapid prototyping • Small dataset • Mutiple backend support	Suitable for: • Flexibility • Short training duration • Debugging capability

This is also known as a front propagated wave which is usually achieved by using a classifying activation function. Unlike in more complex types of neural networks, there is no backpropagation and data move in one direction only. A feedforward neural network may have a single layer or it may have hidden layers. In a feedforward neural network, the sum of the products of the inputs and their weights are calculated. It can be Single-layer perceptron, Multilayer perceptron, or radial basis function [31–40].

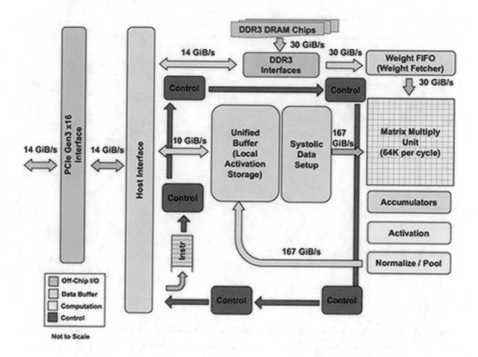

Fig. 4.25 TPU architecture

4.3.1.1 Single-Layer Perceptron(SLP)

It is an algorithm for supervised learning of binary classifiers. A binary classifier is a function which can decide whether or not an input, represented by a vector of numbers, belongs to some specific class. It is a type of linear classifier, i.e., a classification algorithm that makes its predictions based on a linear predictor function combining a set of weights with the feature vector. Considered the first generation of neural networks, **Perceptron** are simply computational models of a single neuron. Also called **feedforward neural network**, perceptron feeds information from the front to the back. Inputs are sent into the neuron, processed, and result in an output. The error being back-propagated is often some variation of the difference between the input and the output. Given that the network has enough hidden neurons, it can theoretically always model the relationship between the input and output (Fig. 4.26). SLP has a limitation when our data is not linear. It does not work fine with multi-classification problems.

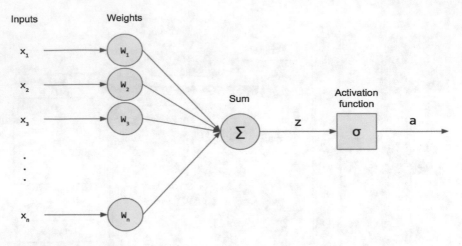

Fig. 4.26 Single-layer perceptron

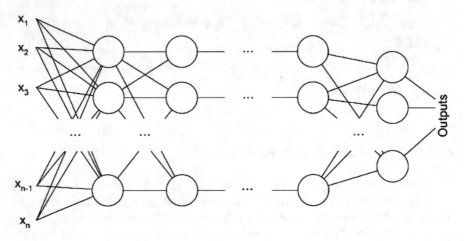

Fig. 4.27 Multilayer perceptron

4.3.1.2   Multilayer Perceptron (MLP)

The limitations of the single-layer network have led to the development of multi-layer feedforward networks with one or more hidden layers, called multilayer perceptron (MLP) networks. A multilayer perceptron has three or more layers. It is used to classify data that cannot be separated linearly. It is a type of artificial neural network that is fully connected. This is because every single node in a layer is connected to each node in the following layer (Fig. 4.27). A multilayer perceptron uses a nonlinear activation function (mainly hyperbolic tangent or logistic function).

4.3.1.3 Radial Basis Function Neural Network

A radial basis function considers the distance of any point relative to the center. Radial basis functions have been applied as a replacement for the sigmoidal hidden layer transfer characteristic in multilayer perceptron. Such neural networks have two layers. In the inner layer, the features are combined with the radial basis function. Then the output of these features is taken into account when calculating the same output in the next time-step. Here is a diagram which represents a radial basis function neural network (Fig. 4.28). The radial basis function neural network is applied extensively in power restoration systems. In recent decades, power systems have become bigger and more complex. This increases the risk of a blackout. This neural network is used in the power restoration systems in order to restore power in the shortest possible time.

4.3.2 Recurrent Neural Network (RNNs)

In the case where you had to memorize previous information traditional neural network approaches including CNN and DNN cannot deal with this due to the following reasons. First, because these approaches only handle fixed size vector as an input and produce a fixed size vector as an output. Second, because those models operate with a fixed number of computational steps (number of layers), so we had to find a new approach to deal with such problem which is RNN (Fig. 4.29).

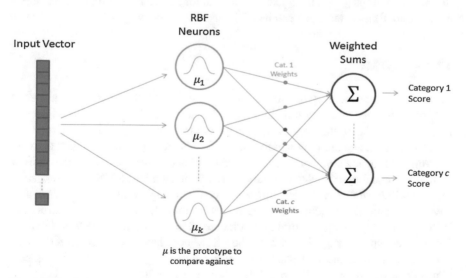

Fig. 4.28 Radial basis function

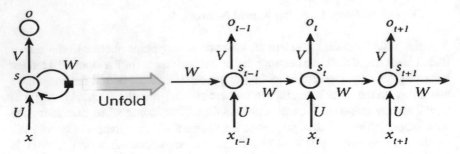

Fig. 4.29 A recurrent neural networks with one hidden unit (left) and its unrolling version in time (right): Temporal

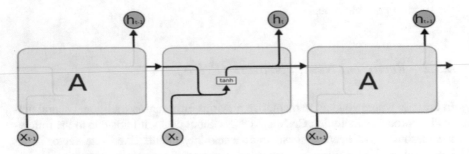

Fig. 4.30 The repeating module in a standard RNN contains a single layer

In recurrent flow, information can flow round in cycles. RNNs are designed to make use of sequential data, when the current step has some kind of relation **with the previous steps** [41]. This makes them ideal for applications with a time component (audio, time-series data) and natural language processing [42–44].

4.3.2.1 LSTMs

Long Short-Term Memory networks—usually just called "LSTMs"—are a special kind of RNN, capable of learning long-term dependencies. They were introduced by Hochreiter and Schmidhuber (1997). They work tremendously well on a large variety of problems, and are now widely used. **LSTMs** are explicitly designed to avoid the long-term dependency problem. All recurrent neural networks have the form of a chain of repeating modules of neural network. In standard RNNs, this repeating module will have a very simple structure, such as a single tanh layer. LSTMs also have this chain like structure, but the repeating module has a different structure. Instead of having a single neural network layer (Fig. 4.30), there are four, interacting in a very special way (Fig. 4.31). The LSTM does have the ability to remove or add information to the cell state, carefully regulated by structures called gates. Gates are a way to optionally let information through. They are composed out of a sigmoid neural net layer and a pointwise multiplication operation. RNNs can be

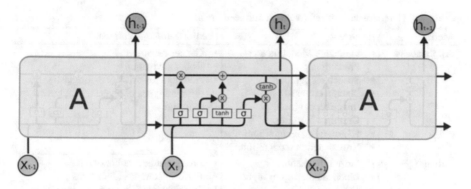

Fig. 4.31 The repeating module in an LSTM contains four interacting layers

used in speech recognition, sentence generation, and language understanding. Comparison between recurrent and feedforward is shown in Table 4.8 [45, 46]. LSTM neural networks can be used to implement various tasks such as prediction, pattern classifications, different types of recognition, analysis, and even sequence generation. Due to capability to process sequential data, LSTM is an efficient tool in many different fields including statistics, linguistics, medicine, transportation, computer science, and others. Different LSTM models are shown in Table 4.9.

4.3.2.2 GRUs

Gated Recurrent Units (GRUs) are LSTMs with different gating. It sounds simple, but lack of output gate makes it easier to repeat the same output for a concrete input multiple times, and are currently used the most in sound (music) and speech synthesis. The actual composition, though, is a bit different: all LSTM gates are combined into so-called update gate, and reset gate is closely tied to input. They are less resource consuming than LSTMs and almost the same effective (Fig. 4.32). GRUs are a gating mechanism in recurrent neural networks, introduced in 2014 by Kyunghyun Cho et al. The GRU is like a long short-term memory (LSTM) with forget gate, but has fewer parameters than LSTM, as it lacks an output gate.

4.3.3 Convolutional Neural Network (CNNs): Feedforward

In feedforward topology, information comes to the input units and flows in one direction through hidden layers until it reaches the output units. Convolutional neural networks, short for "CNN," are a type of feedforward artificial neural networks. Convolutional neural networks are designed for data that comes in the form of multidimensional arrays. Applications include, but are not limited to: image recognition, video understanding, speech recognition, and natural language understanding.

Table 4.8 Comparison between recurrent and feedforward

Metric	Recurrent	Feedforward/convolutional
Applications	• Language Modeling and Prediction. • Speech recognition. • Machine translation. • Image recognition. • Grammar learning. • Handwriting recognition. • Human action recognition.	• Cancer detection • Robot navigation • Autonomous driving: distinguish different types of objects, people, and road signs
Advantages	• Store Information. • Much smaller set of input nodes. • Given an RNN's ability to remember past input, it is suitable for any task where that would be valuable.	• Process huge amount of data • Perform complex algorithms • To achieve the best performance with large amount of data • Feature extraction: take large volumes of data as input, analyze the input to extract features out of an object, and identify similar objects.
Disadvantages	• It cannot process very long sequences if it uses tanh as its activation function. • It is very unstable if we use ReLu as its activation function • RNNs cannot be stacked into very deep models. • RNNs are not able to keep track of long-term dependencies.	• Requires a large amount of data if you only have thousands of example, deep learning is unlikely to outperform other approaches. • Is extremely computationally expensive to train. The most complex models take weeks to train using hundreds of machines equipped with expensive GPUs.
Learning algorithm	• Gradient descent and global optimization methods.	• Supervised learning algorithms: Logistic Regression, Multilayer perceptron, Deep Convolutional Network.
Max # layers	• 3~5	• Max 100

Figure 4.33 shows an example for CNN. CNNs are inspired by how human visual system works [48, 49]. Convolution is a linear operator, dot-product like correlation, not a matrix multiplication, but can be implemented as a sparse matrix multiplication, to be viewed as an affine transform. Convolution **aggregates** information from neighboring pixels. A CNN arranges its neurons in three dimensions (width, height, depth). Every layer of a CNN transforms the 3D input volume to a 3D output volume. In Fig. 4.34, the red input layer holds the image, so its width and height would be the dimensions of the image, and the depth would be 3 (Red, Green, Blue channels). CNN is commonly used in image processing. CNNs consists mainly of four layers: convolutional, activation, pooling, and fully connected layer as depicted in Fig. 4.35. The output feature maps of the final convolution or pooling layer is typically flattened, i.e., transformed into a one-dimensional (1D) array of numbers (or vector), and connected to one or more fully connected layers, also known as dense layers, in which every input is connected to every output by a learnable weight. Once the features extracted by the convolution layers and down-sampled by the pooling layers are

Table 4.9 Different LSTM models [47]

Model	Schematics	Description
One-to-one	y_t ↑ **LSTM** ↑ x_t	One input and one output model is suitable for classifications tasks.
One-to-many	y_t ↑ **LSTM** → y_{t+1} ↑ **LSTM** → ··· y_{t+N} ↑ **LSTM** ↑ x	This model allows to convert a single input to sequences.
Many-to-one	**LSTM** → **LSTM** → ··· **LSTM** ↑ y ↑ x_t x_{t+1} x_{t+N}	Prediction; classification.
Many-to-many	y_t ↑ **LSTM** → y_{t+1} ↑ **LSTM** → ··· y_{t+N} ↑ **LSTM** ↑ x_t x_{t+1} x_{t+N}	Sequence-to-sequence generation (video, text, music); speech recognition; machine translation; video-to-text; image-to-text

Fig. 4.32 GRU

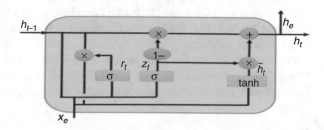

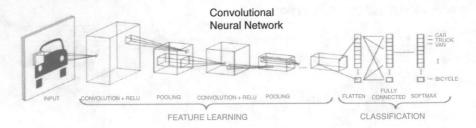

Fig. 4.33 CNN: Spatial

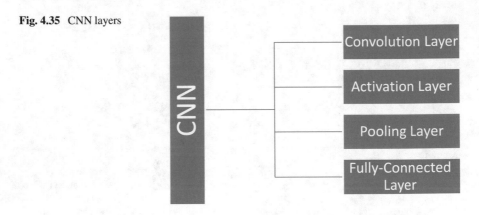

Fig. 4.34 CNN example

Fig. 4.35 CNN layers

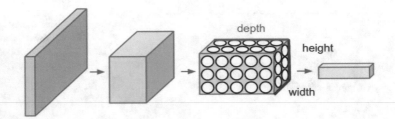

created, they are mapped by a subset of fully connected layers to the final outputs of the network, such as the probabilities for each class in classification tasks. The final fully connected layer typically has the same number of output nodes as the number of classes [50–55]. Each fully connected layer is followed by a nonlinear function, such as ReLU. Pooling can be maximum pooling or average pooling. An example of maximum pooling is shown in Fig. 4.36. There are different architectures for CNN such as *AlexNet* (Fig. 4.37) [56], *VGGNet* (Fig. 4.38) [57], *GoogleNet* (Fig. 4.39) [58], and *ResNet* (Fig. 4.40) [59]. Convolution is an element-wise multiplication. The computer will scan a part of the image, usually with a dimension of 3x3 and multiplies it to a filter, the filter here is the weight. The output of the element-wise multiplication is called a feature map. This step is repeated until all the image is scanned. Note that, after the convolution, the size of the image is reduced.

1 1 2 4
5 6 7 8
3 2 1 0
1 2 3 4
max pool with 2x2 filters
and stride 2
6 8
3 4

Fig. 4.36 An example of maximum pooling

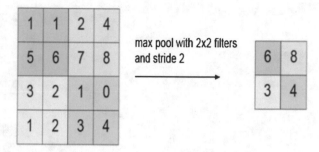

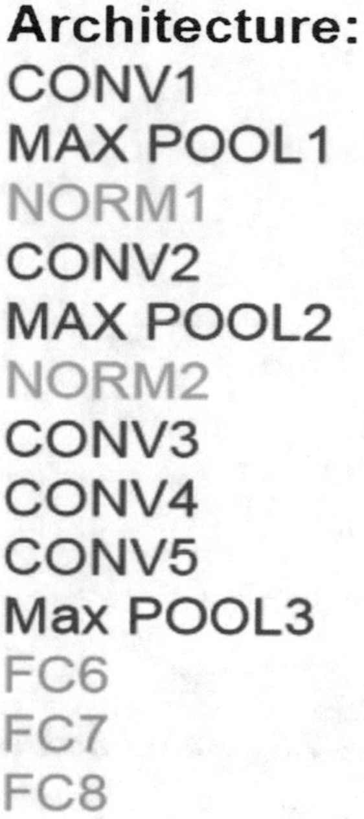

Architecture:
CONV1
MAX POOL1
NORM1
CONV2
MAX POOL2
NORM2
CONV3
CONV4
CONV5
Max POOL3
FC6
FC7
FC8

Fig. 4.37 AlexNet

4.3.4 Generative Adversarial Network (GAN)

GAN is able to create new examples after learning through the real data. It consists of two models competing against each other in a zero-sum game framework. GAN was proposed to generate meaningful images after learning from real photos. It comprises two independent models: the **Generator** and the **Discriminator**. The generator produces fake images and sends the output to the discriminator model. The discriminator works like a judge, as it is optimized for identifying the real photos from the fake ones. The generator model is trying hard to cheat the discriminator while the judge is trying hard not to be cheated. This interesting zero-sum game between these two models motivates both to develop their designed skills and improve their functionalities. Eventually, we take the generator model for producing new images. Figure 4.41 shows GAN functionality.

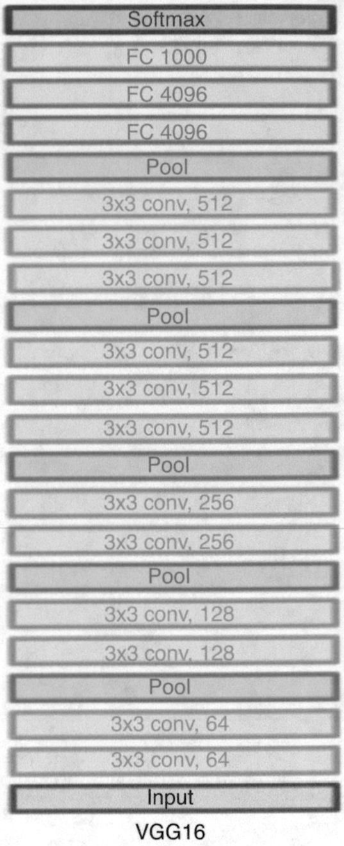

Fig. 4.38 VGGNet

Softmax
FC 1000
FC 4096
FC 4096
Pool
3x3 conv, 512
3x3 conv, 512
3x3 conv, 512
Pool
3x3 conv, 512
3x3 conv, 512
3x3 conv, 512
Pool
3x3 conv, 256
3x3 conv, 256
Pool
3x3 conv, 128
3x3 conv, 128
Pool
3x3 conv, 64
3x3 conv, 64
Input

VGG16

4.3.5   Auto Encoders Neural Network

Autoencoders are used for unsupervised learning. Autoencoders can be used to compress documents on a variety of topics. Autoencoders *encode* input data as vectors. They create a hidden, or compressed, representation of the raw data. They are useful in dimensionality reduction. Figures 4.42 and 4.43 show Autoencoders example for compression. Autoencoders have been applied to many problems and have demonstrated their superiority over classical methods when dealing with noisy or incomplete data. One such application is for data compression. Autoencoders seem to be well suited to this particular function, as they have an ability to preprocess input patterns to produce simpler patterns with fewer components. This compressed information (stored in a hidden layer) preserves the full information obtained from the external environment. The compressed features may then exit the network into the external environment in their original uncompressed form [60].

Autoencoder is a neural network feedforward approach which is trained to predict the input itself in the output. The input layer to hidden layer belongs to the

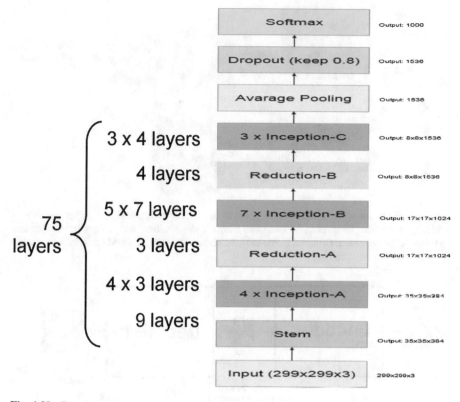

Fig. 4.39 GoogleNet (inception V4)

Fig. 4.40 ResNet

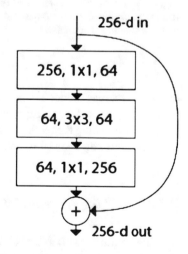

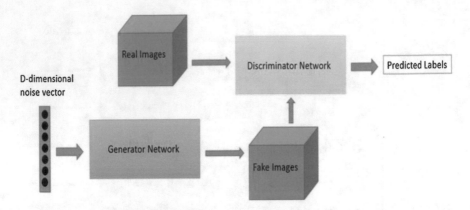

Fig. 4.41 GAN functionality. A discriminator D to classify samples as real or fake. Criminal randomly generates "fake" images. Detective learns to discriminate real images from fakes and becomes better and better. Criminal learns to make better and better fakes. Adversarial relationship helps both of them become better fast

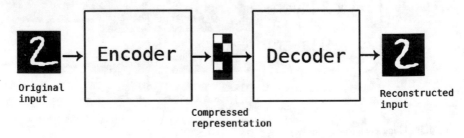

Fig. 4.42 Autoencoder neural networks. Nowadays it is widely used in applications like speech feature extraction and content-based image retrieval. An autoencoder is a fully connected neural network structure that trains a hidden layer as a compressed representation of input

encoder, whereas hidden layer to output layer belongs to the decoder. Autoencoder is an unsupervised learning technique. The optimization of autoencoder is performed by reducing the reconstruction error and learning from the code feature [61]. Autoencoders can be used in data augmentation.

4.3.6 Spiking Neural Network

Spiking neural networks (SNNs) are inspired by information processing in biology, where sparse and asynchronous binary signals are communicated and processed in a massively parallel fashion. SNNs on neuromorphic hardware exhibit favorable properties such as low power consumption, fast inference, and event-driven

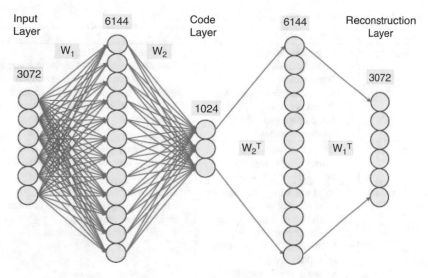

Fig. 4.43 Autoencoder architecture example

information processing. This makes them interesting candidates for the efficient implementation of deep neural networks and the method of choice for many machine learning tasks [62].

Comparison between spike and Convolutional NN is shown in Table 4.10.

4.3.7 Other Types of Neural Network

Symmetrically connected networks (SCNs) are like recurrent networks, but the connections between units are symmetrical, i.e., they have the same weight in both directions. Symmetric networks are much easier to analyze than recurrent networks. They are also more restricted in what they can do because they obey an energy function. SCNs without hidden units are called Hopfield Nets and SCNs with hidden units are called Boltzmann machines.

4.3.7.1 Hopfield Networks

Hopfield networks are trained on a limited set of samples so they respond to a known sample with the same sample. Each cell serves as input cell before training, as hidden cell during training and as output cell when used. As HNs try to reconstruct the trained sample, they can be used for denoising and restoring inputs. Given a half of learned picture or sequence, they will return full sample (Fig. 4.44).

Table 4.10 Comparison between spike and convolutional NN

Metric	Convolutional NN	Spike NN
Applications	• Image recognition. • Video Analysis. • Drug discovery. • Natural language processing. • Checkers game. • Human pose estimation. • Document analysis.	• Information processing. • Study the operation of biological neural circuits, it can model the central nervous system of a virtual insect for seeking food without the prior knowledge of the environment.
Advantages	• More efficient in terms of memory and complexity • Good feature extractors	• Recognizes patterns with little data. • Excellent low resolution performance.
Disadvantages	• From a memory and capacity point of view, the CNN is not much bigger than a regular two-layer network. • At runtime, the convolution operations are computationally expensive and take up about 67% of the time. • CNNs are about 3× slower than their fully connected equivalents (size-wise).	• Due to information load format, the inherent computational costs become much more complicated. • Practical speed limitations.
Learning algorithm	• Supervised training with backpropagation	• One-shot training and feedforward

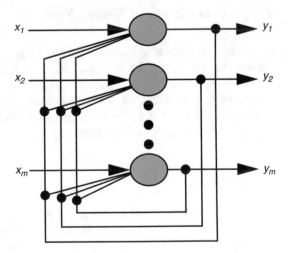

Fig. 4.44 Hopfield network

4.3.7.2   Boltzmann Machine

Boltzmann machines are very similar to HNs where some cells are marked as input and remain hidden. Input cells become output as soon as each hidden cell update their state (during training, BMs / HNs update cells one by one, and not in parallel). This is the first network topology that was successfully trained using **simulated annealing** approach. Multiple stacked Boltzmann Machines can for a so-called Deep belief network (see below), that is used for feature detection and extraction.

4.3.7.3   Restricted Boltzmann Machine

Restricted Boltzmann Machines (RBMs) resembles, in the structure, BMs but, due to being restricted, it is allowed to be trained using backpropagation with the only difference that before backpropagation pass data is passed back to input layer once. It is one type of unsupervised learning neural network [63].

4.3.7.4   Deep Belief Network

Deep belief networks (DBNs) are actually a stack of Boltzmann Machines. They can be chained together when one NN trains another and can be used to generate data by already learned pattern [63].

4.3.7.5   Associative NN

The associative neural network (ASNN) is an extension of committee of machines that combines multiple feedforward neural networks and the k-nearest neighbor technique. It uses the correlation between ensemble responses as a measure of distance amid the analyzed cases for the **kNN**. This corrects the bias of the neural network ensemble. An associative neural network has a memory that can coincide with the training set. If new data become available, the network instantly improves its predictive ability and provides data approximation (self-learns) without retraining. Another important feature of ASNN is the possibility to interpret neural network results by analysis of correlations between data cases in the space of models.

4.4   Challenges for Deep Learning

4.4.1   Overfitting

If you train you network for too long time, the model will start to overfit and learn patterns from the training data that do not generalize to the test data. We have two techniques to overcome the overfitting challenge for CNNs.

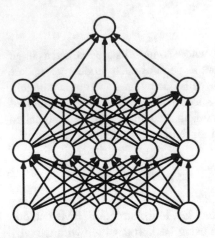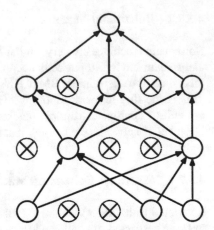

Fig. 4.45 CNN before and after dropout

- **Dropout**: Learning less to learn better. This is done by randomly removing nodes during the training process. An example is shown in Fig. 4.45.
- **Data Augmentation**: Data augmentation means increasing the number of data points. In terms of images, it may mean that increasing the number of images in the dataset. There are many ways to augment data. In images, you can rotate the original image, change lighting conditions, crop it differently, so for one image you can generate different subsamples. Example for data augmentation, where image is exposed to random change in scale, brightness, contrast, saturation and various clipping and flipping is shown in Fig. 4.46.

4.4.2   Underfitting

Underfitting occurs when a statistical model or machine learning algorithm cannot adequately capture the underlying structure of the data. It occurs when the model or algorithm does not fit the data enough. Underfitting occurs if the model or algorithm shows low variance but high bias (to contrast the opposite, overfitting from high variance and low bias). It is often a result of an excessively simple model. Underfitting versus overfitting is shown in Fig. 4.47.

4.5   Advances in Neuromorphic Computing

4.5.1   Transfer Learning

Transfer learning is a machine learning method where a model developed for a task (pretrained models) is reused as the starting point for a model on a second task. Transfer learning is related to problems such as multi-task learning. In transfer

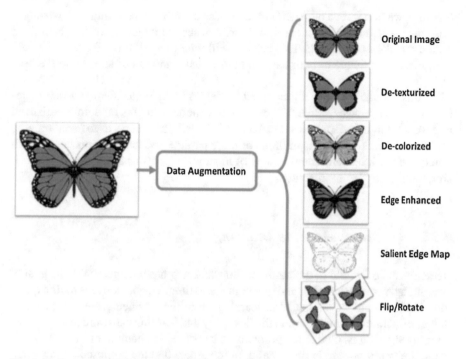

Fig. 4.46 Data augmentation example

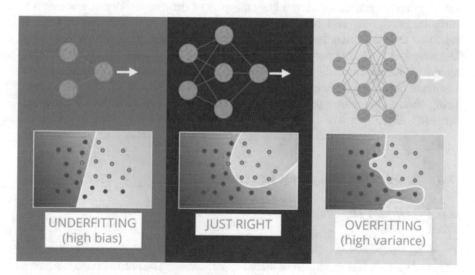

Fig. 4.47 Underfitting versus overfitting [64]

learning, we first train a base network on a base dataset and task, and then we repurpose the learned features, or transfer them, to a second target network to be trained on a target dataset and task. This process will tend to work if the features are general, meaning suitable to both base and target tasks, instead of specific to the base task [65].

Let us assume we have a pretrained model "A" which has learned a rich representation from images, while "B" being a new model which is randomly initialized and we want to transfer knowledge from "A" to "B." This allows us to avoid training "B" from scratch which has millions of more parameters, that requires heavy computational workload and training time of months. "A" acts as a teacher for supervision transfer to "B" architecture.

4.5.2 Quantum Machine Learning

Recently, the new field of quantum machine learning has emerged, with the goal of combining machine learning and quantum computing. The objective is to find quantum implementations of machine learning algorithms which have the expected power of quantum computers and the flexibility and learning capabilities of machine learning algorithms. One of the problems to be solved in quantum machine learning is the limitation present in the quantity of input data that the proposed implementations can handle. Although many-body quantum systems have a Hilbert space whose dimension increases exponentially in relation to the size of the system, permitting to store and manipulate a huge quantity of data, an important problem is to initialize accurately and efficiently this quantum state with the desired data. In machine learning this stage is essential, since learning a problem needs a lot of learning data [66]. Quantum computation promises to expedite the training of classical neural networks and the computation of big data applications, generating faster results. A quron is a qubit in which the two levels stand for active and resting neural firing states. This allows for neural network to be in a superposition of firing patterns [67, 68].

4.6   Applications of Deep Learning

4.6.1   Object Detection

Deep Neural Networks shows an outstanding performance in object detection. Object detection is different from image classification. The entire image is given as input in image classification. But object detection is classifying the localization of objects. First, object locations are computed from the input image. After that, it computes convolutional neural network features and classifies images using output classifier.

We need three basic components to define a basic convolutional network: the convolutional layer, the pooling layer [optional], and the output layer.

Starting with the convolution layer, suppose we have an image of size 6 × 6. We define a weight matrix which extracts certain features from the images as shown in Fig. 4.48 [8]. We have initialized the weight as a 3 × 3 matrix. This weight shall now run across the image such that all the pixels are covered at least once, to give a convolved output. The value 429 in Fig. 4.48 is obtained by adding the values obtained by element-wise multiplication of the weight matrix and the highlighted 3 × 3 part of the input image. The 6 × 6 input images are now converted into an output 4 × 4 image. Convolution operation basically enables parameter sharing in a convolutional neural network.

The weight matrix behaves like a filter in an image extracting particular information from the original image matrix. A weight combination might be extracting edges, while another one might a particular color, while another one might just blur the unwanted noise.

The weights are learnt such that the loss function is minimized. Therefore, weights are learnt to extract features from the original image which help the network in correct prediction. When we have multiple convolutional layers, the initial layer extracts more generic features, while as the network gets deeper, the features extracted by the weight matrices are more and more complex and more suited to the problem at hand. As shown in Fig. 4.48, the filter or the weight matrix was moving across the entire image moving one pixel at a time. We can define it like a hyper parameter, as to how we would want the weight matrix to move across the image. If the weight matrix moves one pixel at a time, we call it as a stride of one. Let us see how a stride of two would look like. The example used in Fig. 4.49 used stride of two.

The second layer is called pooling layer, the input is 4 × 4 convolved image and the Output has become 2 × 2 after the max pooling operation. The max operation is applied to each depth dimension of the convolved output. As shown in Fig. 4.50, the max pooled image still retains the information that it is a car on a street. If you look carefully, the dimensions of the image have been halved. This helps to reduce dimensions. Sometimes when the images are too large, we would need to reduce the

INPUT IMAGE

18	54	51	239	244	188
55	121	75	78	95	88
35	24	204	113	109	221
3	154	104	235	25	130
15	253	225	159	78	233
68	85	180	214	245	0

WEIGHT

Fig. 4.48 The convolution operation and the first output cell from the convolution. The input image has size 6 × 6 and filter applied has size 3 × 3

429	505	686	856
261	782	412	640
633	653	851	751
608	913	713	657

792	856
913	851

Fig. 4.49 The convolved output after using a stride of two of the input image used in Fig. 4.48 on the left, to the right is the output from the maximum pooling

Fig. 4.50 The left the convoluted image and we have applied max pooling on it

number of trainable parameters. It is then desired to periodically introduce pooling layers between subsequent convolution layers. Pooling is done for the sole purpose of reducing the spatial size of the image. Pooling is done independently on each depth dimension; therefore, the depth of the image remains unchanged.

The third layer is called output layer; after multiple layers of convolution and padding, we would need the output in the form of a class. The convolution and pooling layers would only be able to extract features and reduce the number of parameters from the original images. However, to generate the final output we need to apply a fully connected layer to generate an output equal to the number of classes we need. It becomes tough to reach that number with just the convolution layers. Convolution layers generate 3D activation maps while we just need the output as whether or not an image belongs to a particular class. The output layer has a loss function like categorical cross-entropy, to compute the error in prediction. Once the forward pass is complete the back-propagation begins to update the weight and biases for error and loss reduction. Activation function of last layer is different from the previous layers and usually it is softmax function. That is, prior to applying softmax, some vector components could be negative, or greater than one; and might not sum to 1; but after applying softmax, each component will be in the interval {display style $(0,1)$}, and the components will add up to 1, so that they can be interpreted as probabilities. Furthermore, the larger input components will correspond to larger probabilities. Softmax is often used in neural networks, to map the non-normalized output of a network to a probability distribution over predicted output classes. Normalization means changing the scale .We use softmax as the output function of the last layer in neural networks (if the network has n layers, the nth layer is the softmax function). Mathematically the softmax function is shown in Eq. (4.5), where z is a vector of the inputs to the output layer [68].

$$\text{Softmax}(z) = \frac{e^{z_j}}{\sum_{k=1}^{k} e^{z_k}} \tag{4.5}$$

It might be getting a little confusing to understand the input and output dimensions at the end of each convolution layer. We decided to take these few lines to identify the output dimensions. Three hyper parameters would control the size of output volume. The number of filters applied is equal to the depth of the output volume. Remember how we had stacked the output from each filter to form an activation map. The depth of the activation map will be equal to the number of filters.

When we have a stride of one we move across and down a single pixel. With higher stride values, we move large number of pixels at a time and hence produce smaller output volumes. Also, Zero padding helps us to preserve the size of the input image. If a single zero padding is added, a single stride filter movement would retain the size of the original image.

We can apply a simple formula to calculate the output dimensions. The spatial size of the output image can be calculated as ([W-F + 2P]/S) +1. Here, W is the input volume size, F is the size of the filter, P is the number of padding applied, and S is the number of strides. Suppose we have an input image of size $32 \times 32 \times 3$, we apply 10 filters of size $3 \times 3 \times 3$, with single stride and no zero padding. Here W = 32, F = 3, P = 0, and S = 1. The output depth will be equal to the number of filters applied, i.e., 10. The size of the output volume will be ([32–3 + 0]/1) +1 = 30. Therefore, the output volume will be $30 \times 30 \times 10$.

Finally to summarize CNN, it is composed of various convolutional and pooling layers.

First we pass an input image to the first convolutional layer. The convoluted output is obtained as an activation map. The filters applied in the convolution layer extract relevant features from the input image to pass further. Then each filter shall give a different feature to aid the correct class prediction. In case we need to retain the size of the image, we use same padding (zero padding), otherwise valid padding is used since it helps to reduce the number of features. After that the pooling layers are then added to further reduce the number of parameters. Then several convolution and pooling layers are added before the prediction is made. Convolutional layer helps in extracting features. As we go deeper in the network more specific features are extracted as compared to a shallow network where the features extracted are more generic. The output layer in a CNN as mentioned previously is a fully connected layer, where the input from the other layers is flattened and sent so as to transform the output into the number of classes as desired by the network. The output is then generated through the output layer and is compared to the output layer for error generation.

A loss function is defined in the fully connected output layer to compute the mean square loss. The gradient of error is then calculated. The error is then back-propagated to update the filter (weights) and bias values. One training cycle is completed in a single forward and backward pass. However, there is a major problem in CNN. Our main aim is object detection which includes (multi-object detection and classification) the difference between object detection algorithms and classification algorithms is that in detection algorithms, we try to draw a bounding box around the object of interest to locate it within the image. Also, we might not necessarily draw just one bounding box in an object detection case, there could be many bounding boxes representing different objects of interest within the image and we would not know how many beforehand. The major reason why you cannot proceed with this problem by building a standard convolutional network followed by fully connected layer is that, the length of the output layer is variable—not constant, this is because the number of occurrences of the objects of interest is not fixed [69–71].

4.6.2   Visual Tracking

Visual Tracking is the automatic computation of the path of an object which moves from one place to another in a video. It has various applications which include security, analysis of sports video or human–computer interaction. When a particular problem needs multiple objects to be tracked while moving, proper setting is to be done for each object individually. Once object tracking is identified in the first frame of video, subsequent frames of an object are easy to identify. It is a very challenging problem in IOT analytics [72].

4.6.3   Natural Language Processing

Natural language processing (NLP), which aims to enable computers to process human languages intelligently, is an important interdisciplinary field crossing artificial intelligence, computing science, cognitive science, information processing, and linguistics. Concerned with interactions between computers and human languages, NLP applications such as speech recognition, dialog systems, information retrieval, question answering, and machine translation have started to reshape the way people identify, obtain, and make use of information. Neural machine translation has quickly become the new de facto technology in major commercial online translation services offered by large technology companies: Google, Microsoft, Facebook, Baidu, and more [73].

4.6.4 Digits Recognition

The **input layer** of the network contains neurons encoding the values of the input pixels. Assume our training data for the network will consist of many 16 by 16 pixel images of scanned handwritten digits, and so the input layer contains $256 = 16 \times 16$ neurons. The input pixels are grayscale, with a value of 0.0 representing white, a value of 1.0 representing black, and in between values representing gradually darkening shades of gray.

The second layer of the network is a hidden layer. We denote the number of neurons in this hidden layer by n, and the example illustrates a small hidden layer, containing just $n = 15$ neurons.

The output layer of the network contains 10 neurons. If the first neuron fires, i.e., has an output ≈ 1, then that will indicate that the network thinks the digit is a 0. If the second neuron fires, then that will indicate that the network thinks the digit is a 1. And so on. A little more precisely, we number the output neurons from 0 through 9, and figure out which neuron has the highest activation value. If that neuron is, say, neuron number 6, then our network will guess that the input digit was a 6. And so on.

4.6.5 Emotions Recognition

Facial expressions convey emotions and provide evidence on the personalities and intentions of people. Studying and understanding facial expressions has been a long-standing problem. The first reported scientific research on the analysis of facial expressions can be traced back to as early as 1862 to Duchenne who wanted to determine how the muscles in the human face produce facial expressions. Charles Darwin also studied facial expressions and body gestures in mammals. An influential milestone in the analysis of facial expressions was the work of Paul Ekman, who described a set of six basic emotions: anger, fear, disgust, happiness, sadness, and surprise. They are universal in terms of expressing, and understanding them.

4.6.6 Gesture Recognition

Hand gesture recognition system received great attention in the recent few years because of its manifoldness applications and the ability to interact with machine efficiently through human–computer interaction. The essential aim of building hand gesture recognition system is to create a natural interaction between human and computer where the recognized gestures can be used for controlling a robot or conveying meaningful information. How to form the resulted hand gestures to be understood and well interpreted by the computer considered as the problem of gesture interaction. Machine learning is very promising in gesture recognition.

4.6.7 Machine Learning for Communications

ML is conventionally thought to have its application justified in the situations, where there is no exact mathematical model of the system available, a sufficiently large amount of training data is available, the system/model under study is stationary along time, and the numerical analysis is acceptable. The ML techniques have recently gained significant attention for the provision of data-driven solutions to various challenging problems in communication systems. The deployment of ML in communications is rapidly gaining popularity, in particular, to build self-sustaining and adaptive networks capable to meet the dynamic reconfigurability demands of the future devices and services. Furthermore, ML has a strong potential to replace the conventional mathematical model-based algorithmic solutions, given the availability of adequate data and computational power.

4.7 Cognitive Computing: An Introduction

A cognitive system accumulates a considerable amount of knowledge to give fact-based advice to both customers and employees. It is a self-learning system that uses **data mining, pattern recognition, and natural language processing** to replicate the way the human brain works. It inputs a combination of structured and unstructured data, both external and internal, to process any decision-making. The more the system learns, the more efficient and accurate it becomes, leading to a virtuous cycle of efficiency and customer satisfaction.

A cognitive computing mimics **human reasoning** methodologies, by exploiting the accumulated experience to learn from the past, errors, and successful findings.

Cognitive computing may need hardware that supports fine grained parallelism like the brain, with distributes memory and computation as the brain is massively parallel and is massively interconnected.

The seven-layer Layered Reference Model of the Brain (**LRMB**) model can be refined by 43 cognitive processes as shown in Fig. 4.51. Applications of cognitive computing include search engines, healthcare, and education.

In cognitive computing, new hardware or software devices mimic human brain and take a decision appropriate to the situation. Moreover, cognitive computing is used in numerous artificial intelligence (AI) applications, including expert systems, natural language programming, neural networks, robotics, and virtual reality. Self-learning capability of human beings is adapted to the system by applying artificial intelligence to it.

Cognitive systems are not programmed in prior, rather they are designed to augment themselves by learning through training, interactions, previous experience, and past reference datasets. Thus, in contrast to conventional programmable (Von-Neumann) computers, Cognitive Computing does not limit itself within the

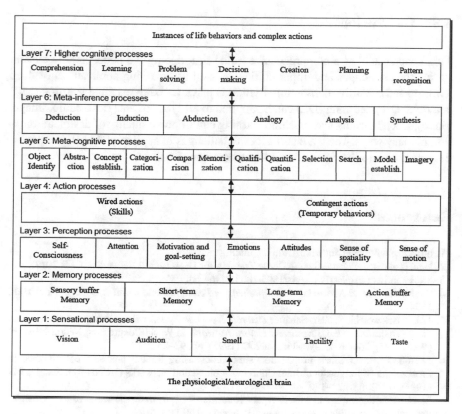

Fig. 4.51 The seven-layer LRMB model [74]

deterministic boundaries. It brings a dynamic essence to the systems by continuously sensing and learning from the surroundings and enhancing its decision-making capabilities.

Cognitive Computing is the blend of different capabilities as ML, NLP, cognitive vision, reasoning and learning, etc. Leveraging these capabilities, Cognitive Computing unlocks the information from massive data to develop deep and predictive insights. The output of Cognitive Computing may be prescriptive, suggestive, or instructive. By exploiting past errors and success, cognitive systems help machines to learn and teach humans new concepts and/or behaviors. The unique combination of analytics, problem-solving, and communication with a human in the natural form redefines the relationship between human and machines, whereby machines augment human in decision support and reasoning [75–77].

4.8 Conclusions

This chapter acts as a comprehensive guide introducing the concepts, the current status, the challenges and opportunities, as well as emerging applications of deep learning and cognitive computing. Python is just a tool, but machine learning (ML) is a field. Deep learning (DL) can be applied to all types of data (test, audio, image, video). Deep learning and machine learning can improve quality of design automation in many ways. There are still a lot of challenges in applying DL. ML is a potential alternative to the conventional logic-based approaches.

References

1. https://en.wikipedia.org/wiki/Artificial_intelligence
2. Han, et al., DSD: Dense-Sparse-Dense Training for Deep Neural Networks (ICLR 2017)
3. https://arxiv.org/ftp/arxiv/papers/1803/1803.01164.pdf
4. S. Haykin, *Neural Networks: A Comprehensive Foundation* (Macmillan College, New York, 1994)
5. J.G. Kusckewschi, S. Zein-Sabatto, *Discrete Dynamic Systems Parameter Identification and State Estimation Using a Recurrent Neural Network*, in World Congress on Neural Network (WCNN'95) Conference, Washington DC (1995), pp. 92–95
6. J. Holt, T. Baker, Back propagation simulations using limited precision calculations, in Proceedings of International Joint Conference on Neural Networks (IJCNN-91), Seattle, WA (1991), vol. 2, pp. 121–126
7. Stackoverflow. https://stackoverflow.com/questions/4674623/why-do-we-have-to-normalize-the-input-for-an-artificial-neural-network. Accessed 29 Dec 2016
8. T. O'Shea, J. Hoydis, An introduction to deep learning for the physical layer. IEEE Trans. Cogn. Commun. Netw. **3**(4), 563–575 (2017)
9. K. Arulkumaran, M.P. Deisenroth, M. Brundage, A.A. Bharath, Deep reinforcement learning: A brief survey. IEEE Signal Process. Mag. **34**(6), 26–38 (2017)
10. A. P. Canziani, E. Culurciello, *An Analysis of Deep Neural Network Models for Practical Applications* (CoRR, abs/1605.07678, 2016)
11. E. Nurvitadhi et al., *Can FPGAs Beat GPUs in Accelerating Next-Generation Deep Neural Networks?* In Proc. 2017 ACM/SIGSA International Symposium of Field-Programmable Gate Arrays 5–14 (Association for Computing Machinery, 2017)
12. S. Rashchka, Python Machine Learning: Unlock deeper insights into machine learning with this vital guide to cutting-edge predictive analytics (Packt, Birmingham 2016) 2nd Ed. ISBN 978178355-513-0
13. S. Shalev-Shwartz, S. Ben-David, *Understanding Machine Learning: From Theory to Algorithms*, 7th edn. (Cambridge University Press, Cambridge, 2017). ISBN 978-1-107-05713-5
14. D.L. Poole, A.K. Mackworth, *Artificial Intelligence: Foundations of Computational Agents*, 2nd edn. (Cambridge University Press, Cambridge, 2017). ISBN 978-1-107-19539-4
15. https://caffe2.ai/
16. https://www.tensorflow.org/
17. http://torch.ch/
18. https://pytorch.org
19. https://docs.microsoft.com/en-us/cognitive-toolkit/
20. https://mxnet.apache.org/

21. https://keras.io/. Accessed June 2019
22. https://opencv.org/
23. https://towardsdatascience.com/the-mostly-complete-chart-of-neural-networks-explained-3fb6f2367464
24. https://www.digitalvidya.com/blog/types-of-neural-networks/
25. https://pytorch.org/
26. N.P. Jouppi, C.U.A. Young, *In-Datacenter Performance Analysis of a Tensor Processing Unit* (Google, Inc, Mountain View, 2017)
27. https://www.nvidia.com/en-gb/data-center/volta-gpu-architecture/
28. https://www.microsoft.com/en-us/research/project/project-brainwave/
29. http://www.myhdl.org/
30. http://www.think-force.com/en/site/chip
31. S. Han, et al., *EIE: Efficient Inference Engine Oncompressed Deep Neural Network* (ISCA, 2016)
32. A. Krizhevsky et al., *Imagenet Classification with Deep Convolutional Neural Networks* (NIPS, Long Beach, 2012)
33. A. Graves et al., *Speech Recognition with Deep Recurrent Neural Net-Works* (ICASSP, Barcelona, 2013)
34. L. Yavits et al., Computer architecture with associative processor replac-ing last-level cache and SIMD accelerator. IEEE Trans. Comput. **64**(2), 368–381 (2014)
35. L. Yavits et al., Sparse matrix multiplication on an associative processor. IEEE Transactions on Parallel and Distributed Systems **26**(11), 3175–3183
36. D. Shin, et al., *DNPU: An 8.1TOPS/W Reconfigurable CNN-RNNPro-cessor for General-Purpose Deep Neural Networks* (ISSCC, 2017)
37. F. Conti, et al., *Chipmunk: A Systolicallyscalable 0.9 mm 2, 3.08 Gop/s/mW @ 1.2 mW Accelerator for Near-Sensor Recurrent Neural Net-work Inference* (IEEE CICC, 2018)
38. V. Rybalkin, et al, *Hardware Architecture of BidirectionalLong Short-Term Memory Neural Network for Optical Character Recognition* (IEEEDATE, 2017), pp. 1390–1395
39. Y. Chen et al., *DaDianNao: A Machine-Learning Supercomputer* (MI-CRO 2014)
40. P. Chi, et al., *PRIME: A Novel Processing-in-Memory Architecture for Neural Network Computation in ReRAM-Based Main Memory* (ISCA, 2016)
41. R. Pascanu, T. Mikolov, Y. Bengio, *On the Difficulty of Training Recurrent Neural Networks.* In Proceedings of the 30th International Conference on Machine Learning (2013)
42. L. Deng, O. Abdel-Hamid, D. Yu, *A Deep Convolutional Neural Network Using Heterogeneous Pooling for Trading Acoustic Invariance with Phonetic Confusion.* In Proceedings of the 2013 IEEE International Conference on Acoustics, Speech and Signal Processing (ICASSP), Vancouver, BC, Canada (2013), pp. 6669–6673
43. A. Graves, A.-R. Mohamed, G. Hinton, *Speech Recognition with Deep Recurrent Neural Networks.* In Proceedings of the 2013 IEEE International Conference on Acoustics, Speech and Signal Processing (ICASSP), Vancouver, BC, Canada (2013), pp. 6645–6649
44. L. Deng, D. Yu, Deep learning: Methods and applications. Found. Trends Signal Process. **7**, 197–387 (2014)
45. L. Deng, X. Li, Machine learning paradigms in speech recognition: An overview. IEEE Trans. Audio, Speech, & Language (2013)
46. G. Hinton, N. Srivastava, A. Krizhevsky, I. Sutskever, R. Salakhutdinov, *Improving Neural Networks by Preventing Co-adaptation of Feature Detectors*, arXiv: 1207.0580v1 (2012)
47. Z. Zhao, A. Srivastava, L. Peng, Q. Chen, Long short-term memory network design for analog computing. ACM Journal on Emerging Technologies in Computing Systems (JETC) **15**(1), 13 (2019)
48. W. Choi et al., On-Chip communication network for efficient training of deep convolutional networks on heterogeneous Manycore systems. IEEE TC **67**(5), 672–686 (2018)
49. R.G. Kim et al., *Machine Learning and Manycore Systems Design: A Serendipitous Symbiosis* (IEEE Computer, 2018)

50. S. Gupta, A. Agrawal, K. Gopalakrishnan, et al., *Deep Learning with Limited Numerical Precision*. In: Proceedings of the 32nd International Conference on International Conference on Machine Learning, vol. 37 (ICML'15; 2015), pp. 1737–1746

51. T. Dettmers, *8-Bit Approximations for Parallelism in Deep Learning. Computing Research Repository* (2015). abs/1511.04561. http://arxiv.org/abs/1511.04561

52. M. Courbariaux, Y.Bengio, J.P.David, *BinaryConnect: Training Deep Neural Networks with Binary Weights During Propagations*. In: Proceedings of the 28th International Conference on Neural Information Processing Systems, vol. 2 (NIPS'15, 2015), pp. 3123–3131

53. M. Courbariaux, Y. Bengio, *BinaryNet: Training Deep Neural Networks with Weights and Activations Constrained to +1 or −1*. Computing Research Repository (2016). abs/1602.02830. http://arxiv.org/ abs/1602.02830

54. R. Zhao, W. Song, W. Zhang, et al., *Accelerating Binarized Convolutional Neural Networks with Software-Programmable FPGAs*. In: Proceedings of the 2017 ACM/SIGDA International Symposium on Field-Programmable Gate Arrays, (FPGA'17, 2017), pp. 15–24

55. Y. Umuroglu, N.J. Fraser, G. Gambardella, et al. *FINN: A Framework for Fast, Scalable Binarized Neural Network Inference*. In: Proceedings of the 2017 ACM/SIGDA International Symposium on Field-Programmable Gate Arrays (FPGA'17, 2017), pp. 65–74

56. A. Krizhevsky, I. Sutskever, G. E. Hinton, *ImageNet Classification with Deep Convolutional Neural Networks*, In: Neural Information Processing Systems (NIPS) (2012)

57. K. Simonyan, A. Zisserman, *Very Deep Convolutional Networks for Large-Scale Image Recognition*, In: International Conference on Learning Representations (ICLR) (2015)

58. C. Szegedy et al., *Going Deeper with Convolutions*, In: IEEE Conference on Computer Vision and Pattern Recognition (CVPR) (2015)

59. K. He et al., Deep Residual Learning for Image Recognition, In: IEEE Conference on Computer Vision and Pattern Recognition (CVPR) (2016)

60. G. E. H., R. R. S., Reducing the dimensionality of data with neural networks. Science **313**, 504–507 (2006)

61. C.Y. Liou, W.C. Cheng, J.W. Liou, et al., Autoencoder for words. Neurocomputing **139**, 84–96 (2014)

62. https://www.frontiersin.org/articles/10.3389/fnins.2018.00774/full

63. Y. Ren, Y. Wu, *Convolutional Deep Belief Networks for Feature Extraction of EEG Signal*, 2014 Int. Jt. Conf. Neural Networks (2014), pp. 2850–2853

64. https://www.tensorflow.org/tutorials/keras/overfit_and_underfit

65. https://machinelearningmastery.com/transfer-learning-for-deep-learning/

66. M. Schuld, I. Sinayskiy, F. Petruccione, An introduction to quantum machine learning. Contemp. Phys. **56**, 172 (2015)

67. A. Narayanan, T. Menneer, Quantum artificial neural network architectures and components. Inf. Sci. **128**(3–4), 231–255 (2000)

68. https://www.mikulskibartosz.name/understanding-the-softmax-activation-function/

69. S. Han, J. Pool, J. Tran, et al., *Learning both Weights and Connections for Efficient Neural Network*. In: Advances in Neural Information Processing Systems 28: Annual Conference on Neural Information Processing Systems 2015, Montreal, Quebec, Canada, 7–12 December 2015 (2015), pp. 1135–1143

70. LeCun Y, J. S. Denker, S. A. Solla. *Optimal Brain Damage*. Advances in Neural Information Processing Systems 2 [NIPS Conference, Denver, Colorado, USA, November 27–30, 1989] (Morgan Kaufmann, 1990), pp. 598–605

71. S. Anwar, K. Hwang, W. Sung, Structured pruning of deep convolutional neural networks. ACM J. Emerg. Technol. Comput. Syst. **13**(3), 32:1–32:18 (2017)

72. A.K. Sangaiah, A. Thangavelu, V.M. Sundaram, *Cognitive Computing for Big Data Systems over IoT* (Springer, Berlin, 2018)

73. L. Deng, Y. Liu, *Deep Learning in Natural Language Processing* (Springer, Berlin, 2019)

74. Y. Wang, On cognitive computing. Int. J. of Software Science and Computational Intelligence **1**(3), 1–15 (2009)

75. E. John, I.I.I. Kelly, *Steve Hamm:Smart Machines—IBM's Watson and the Era of Cognitive Computing* (Columbia University Press, New York, 2013). ISBN 978-0-231-16856-4
76. E. Brynjolfsson, *Andrew McAfee:The SecondMachine Age –Work, Progress, and Prosperity in a Time of Brilliant Technologies* (W.W. Norton, New York, 2014). ISBN 978-0-393-23935-5
77. W. Eric Brown, *Cognitive Computing Ushers in New Era of IT* (2014). http://www.forbes.com/sites/ibm/2014/02/03/cognitive-computing-ushers-in-new-era-of-it/. Accessed 8 Sep 2014

Chapter 5
Approximate Computing: Towards Ultra-Low-Power Systems Design

5.1 Introduction

Power consumption has emerged to be a major design constraint for modern integrated circuits, especially for low power applications such as the "Internet of Things," wearable, and implantable devices. Approximate computing (AC) is a promising paradigm to overcome the energy scaling barrier of computer systems. Approximate computing is a special type of computation which returns a possibly inaccurate result—but acceptable—rather than a guaranteed accurate result to save resources, memory, run-time, and energy. The motivation behind this is that some applications are computing their results more accurately than needed, so they waste resources. So, it trades off between accuracy in computation or acceptable quality of results and resources [1]

Approximate computing should be applied only to noncritical data applications and error-resilient systems [2–4]. Some applications cannot be approximated such as cryptography. Approximate computing is computing like a human. Approximate computing has variety of applications such as machine learning, computer vision, recognition, mining, big data analytics, signal processing, multimedia processing, and Internet of Things (IoT) [5–7]. The quality metrics in AC is the relative error from standard output and the target output quality (TOQ) requirements or the approximation error [8]. The approximate computing dimensions are resource usage, performance, and accuracy as depicted in Fig. 5.1. In this chapter, a survey of recent techniques for AC is presented [9, 10]

AC allows leveraging the inherent tolerance of some applications to imprecisions in order to achieve higher efficiencies and performance gains. Some of these applications include computer vision/graphics, image processing, and digital signal processing (DSP).

The golden rule in approximate computing is "don't compute more than needed." Approximate computing can be performed using several techniques on several

© Springer Nature Switzerland AG 2020
K. S. Mohamed, *Neuromorphic Computing and Beyond*,
https://doi.org/10.1007/978-3-030-37224-8_5

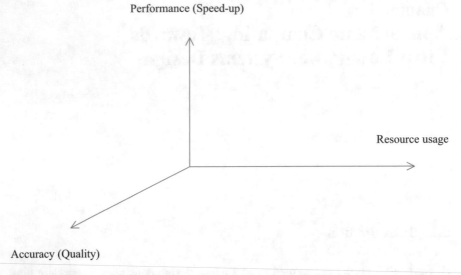

Fig. 5.1 Dimensions of approximate computing. Approximate Computing has two goals: to improve resource consumption at the expenses of accuracy and do so while maintaining programs' Quality of Service (QoS)

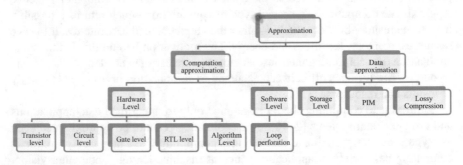

Fig. 5.2 Classification of approximation

dimensions "approximate computing on approximate data" as depicted in Fig. 5.2 [11–14].

- **Hardware-level approximation**: approximate arithmetic and logical **circuits** such as adders, multipliers, and XOR operation can reduce hardware overhead. Moreover, high-level synthesis (HLS) approximates some functions.
- **Software-level approximation**: Approximate operations such as skipping some iteration of loops, skipping some tasks, reducing the precision, and reducing the width of the data bits.
- **Data-level approximation**: **store** data values approximately, reduce data movement or **communication** using processing in memory (**PIM**) computing to move computation into memory, use lossy compression.

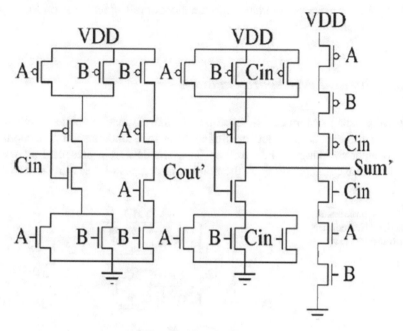

Fig. 5.3 Accurate full adder which consists of 24 transistors [19]

5.2 Hardware-Level Approximation Techniques

Addition and multiplication are the two basic operations in applications like signal and media processing and are key components in any logic circuit. Approximations can be done on transistor level, gate level, RTL level, or algorithmic level. So, all levels of design abstraction can be subject to approximations provided the output quality is met. Examples are as follows:

- Transistor level: adders [15].
- Gate level: multipliers [16].
- RTL level: image filters, iterative algorithms [17].
- Algorithm level: neural network (NN) [18].
- Device level: Memristor-Based Approximate Matrix Multiplier.

5.2.1 Transistor-Level Approximations

There are many works on approximating full adder cells with reduced complexity at the transistor level. An example for full adder which consists of 24 transistors is shown in Fig. 5.3. While an approximated version is shown in Fig. 5.4, which consists of 11 transistors. Another approximation for full adder based on XOR function is shown in Fig. 5.5.

5.2.2   Circuit-Level Approximations

Any real physical system can be substituted by a mathematical model. Modeling is important for explaining data, predicting behaviors, understanding phenomena, optimizing or checking, and controlling. These models are represented by a set of equations. For complex systems, the number of degrees of freedom (independent

Fig. 5.4 Approximated full adder which consists of 11 transistors [19]

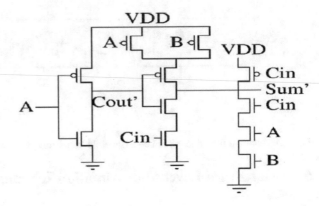

Fig. 5.5 Approximated full adder which consists of 8 transistors [19]

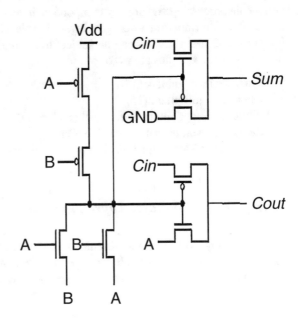

variables) is often large and thus determination of a solution can be hard and computationally expensive either in terms of long CPU time or heavy memory storage. Model order reduction is widely used to approximate a high order model by a reduced order one. The reduced model should still acquire the main characteristics of original system. Thus, the time needed to solve the new set of equations is significantly reduced. Circuit simulation is an imported example for model order reduction deployment. Consider, a tenth order transfer function expressed as follows:

$$G_1(s) = \frac{\begin{array}{c} s^9 + 46.8s^8 + 957.6s^7 + 11144s^6 + 80511.9s^5 \\ +369601.6s^4 + 1060774.5s^3 + 1809006.4s^2 + 1669955.4s + 638266 \end{array}}{\begin{array}{c} s^{10} + 36.9s^9 + 620.8s^8 + 6257.9s^7 + 41888s^6 + 195879.7s^5 \\ +658023.2s^4 + 1611073.5s^3 + 2857356s^2 + 3425885.4s + 2110138.4 \end{array}} \quad (5.1)$$

The reduced second order model is expressed as follows:

$$R_1(s) = \frac{2.574s + 1.847}{s^2 + 0.707s + 6.238} \quad (5.2)$$

5.2.3 Gate-Level Approximations

5.2.3.1 Approximate Multiplier Using Approximate Computing

An example of approximate computing is to approximate the output of 2-bit multiplier to be represented by 3 bits instead of 4 bits as depicted in Fig. 5.6. This approximation reduces both the area and the delay of the original multiplier significantly. Compared with an accurate multiplier, the proposed multiplier can consume 70% less power with average computational error <2%.

5.2.3.2 Approximate Multiplier Using Stochastic/Probabilistic Computing

Traditional arithmetic circuits represent numerical values with zeros and ones. Stochastic encoding is another way to represent numerical values, where a real value p in the unit interval is represented by a sequence of N *random* bits X_1, X_2, ..., X_N, with each X_i having probability p of being one and probability $(1-p)$ of being zero, i.e., $P(X_i = 1) = p$ and $P(X_i = 0) = 1-p$. Probability means "the chance that something will happen." An example for stochastic encoding is shown in Fig. 5.7 [20–24]. Stochastic numbers (SNs) are interpreted as probabilities in the [0, 1] interval.

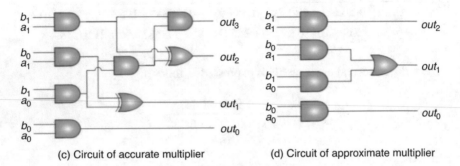

(a) Karnaugh map of accurate multiplier   (b) Karnaugh map of approximate multiplier

(c) Circuit of accurate multiplier   (d) Circuit of approximate multiplier

Fig. 5.6 Accurate and approximate designs of 2-bit multiplier [3]. If we change the output "1001" in the red circle in the Karnaugh map of an accurate 2-bit multiplier to "111," we obtain the Karnaugh map of an approximate 2-bit multiplier [16]

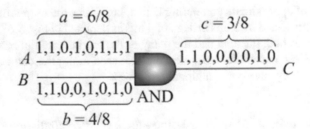

Fig. 5.7 An AND gate multiplying two input stochastic bit streams. A is 6/8 because total number of ones are 6 and total number of bits are 8. B is 4/8 because total number of ones are 4 and total number of bits are 8. C is the adding of stochastic encoded A, B which is equivalent to multiplying them in binary encoding [4]

5.2.4   RTL-Level Approximations

5.2.4.1   Iterative Algorithms

Iterative algorithms are inherently resilient to approximation because errors introduced in one iteration are fixed in later iterations. Iterative algorithms can be found for many numerical analysis problems encountered in scientific computing. The

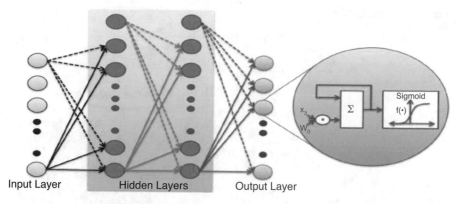

Fig. 5.8 A simplified diagram of neural network

Conjugate Gradient method used to solve systems of linear equations performs well using low precision arithmetic for the computationally expensive parts [17].

5.2.5   Algorithm-Level Approximations

5.2.5.1   Iterative Algorithms

Neural networks (NN) are used to implement multiplications. The general architecture of NN is shown in Fig. 5.8 [7]. The idea is to use machine learning to imitate a computation by observing its input/output behavior. Then, execute the learned model in place of the original code. Other algorithms can be used in approximate computing such as genetic algorithm. Approximate Coordinate Rotation Digital Computer (CORDIC) implementation is another example for approximate computing at algorithm level [25]. The CORDIC algorithm is a clever method for accurately computing trigonometric functions using only additions, bit shifts, and a small lookup table. Moreover, NN can be used to implement ALU or basic gates [26–28].

5.2.5.2   High-Level Synthesis (HLS) Approximations

HLS is a process, which takes as inputs an untimed behavioral description and generates efficient RTL code (Verilog or VHDL), which can execute it by performing three main steps: (1) resource allocation; (2) scheduling; and (3) binding as depicted in Fig. 5.9. Many floating point arithmetic can be approximated by HLS [29–31].

- **Allocation**: It defines the type and the number of hardware resources (for instance, functional units, storage, or connectivity components) needed to satisfy

Fig. 5.9 Design steps for
HLS

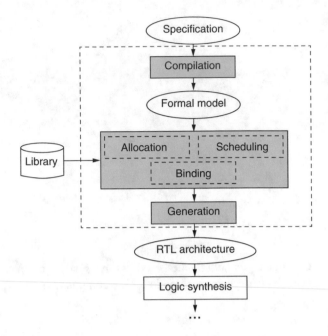

the design constraints. Depending on the HLS tool, some components may be added during scheduling and binding tasks. For example, the connectivity components (such as buses or point-to-point connections among components) can be added before or after binding and scheduling tasks. The components are selected from the RTL component library. It is important to select at least one component for each operation in the specification model. The library must also include component characteristics (such as area, delay, and power) and its metrics to be used by other synthesis tasks [32].

- **Scheduling**: All operations required in the specification model must be scheduled into cycles. Depending on the functional component to which the operation is mapped, the operation can be scheduled within one clock cycle or scheduled over several cycles. Operations can be chained (the output of an operation directly feeds an input of another operation). Operations can be scheduled to execute in parallel provided there are no data dependencies between them and there are sufficient resources available at the same time [32].

- **Binding:** Each variable that carries values across cycles must be bound to a storage unit. In addition, several variables with nonoverlapping or mutually exclusive lifetimes can be bound to the same storage units. Every operation in the specification model must be bound to one of the functional units capable of executing the operation. If there are several units with such capability, the binding algorithm must optimize this selection. Storage and functional unit binding also depend on connectivity binding, which requires that each transfer from component to component be bound to a connection unit such as a bus or a multiplexer [32].

5.2.6   Device-Level Approximations: Memristor-Based Approximate Matrix Multiplier

Although analog computers sound attractive alternatives to digital processor, they have become extinct during the past decades because of their drawbacks. Fortunately, some changes are happening. The first drawback of analog computing is its low accuracy, which is now tolerable by introducing approximate computing. Lack of effective analog memories and non-configurability are the other major problems before analog computing. It seems that memristors can fulfill this requirement. Memristors in the field of analog computing can be categorized into two main approaches. The first approach is to use memristors as synapses in artificial neural networks. The second approach is to develop memristor crossbar-based analog computational blocks often for the multiplication and digital signal processing. In this approach, the memristor crossbars usually play the role of vector–matrix multiplier for implementing the desired algorithms as shown in Fig. 5.10 [33].

5.3   Software-Level Approximation Techniques

5.3.1   Loop Perforation

Some programs consume too much energy, one solution to overcome this is to perforate the loops, i.e., do not execute all loop iterations. Instead, skip some iteration as illustrated in Fig. 5.11. Loop perforation typically makes computations run faster by doing less work.

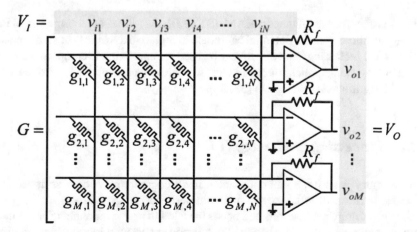

Fig. 5.10 Memristor crossbars for vector–matrix multiplication

for (i = 0; i < n; i++) { ...} $\implies$ for (i = 0; i < n; i=i+2) { ...}

Fig. 5.11 Loop perforation example

5.3.2 Precision Scaling

Several architectures work by changing the precision (bit width) of input to reduce computing requirements. Dynamic precision scaling is used for improving efficiency of physics-based animation. This technique finds the minimum precision required by Switch between different fixed-point specifications and performing profiling at the design time. At runtime, the energy difference between consecutive simulation steps is measured and compared with a threshold to detect whether the simulation is becoming unstable. In case of instability, the precision is restored to the maximum, and as simulation stabilizes, the precision is progressively reduced until it reaches the minimum value.

5.3.3 Synchronization Elision

It refers to a relaxation in the synchronization in parallel applications to improve performance [34]

5.4 Data-Level Approximation Techniques

5.4.1 STT-MRAM

Spin Transfer Torque Magnetic RAM (STT-MRAM) is an emerging nonvolatile memory technology and a potential candidate to replace SRAM in processor caches. STT-MRAM technology can effectively be used for approximate computing by tuning technology and application parameters to achieve an acceptable level of correctness with significant power gains [35, 36].

5.4.2 Processing in Memory (PIM)

Main memory is an important component in all computing systems, so it needs to be scaled in terms of efficiency to maintain performance growth and scaling pros. Recent technology applications and needs have led to new requirement for the main memory (more capacity, more bandwidth, less power consumption, and computations near memory as data movement is much more energy-hungry than computation). DRAM and flash memories do not satisfy all these requirements. There is a memory capacity gap with core count doubling every 2 years and DRAM capacity doubling every 3 years.

Besides, DRAM consumes power even when not used. Moreover, according to ITRS, Scaling DRAM beyond 40–35 nm is challenging. Also, there is a problem with some recent DRAM devices in which repeatedly accessing a row of memory can cause bit flips in adjacent rows [37]. So, there is still a room for improvement in the current technology or development of emerging technologies.

Processing-in-memory (PIM) provides high bandwidth and high energy efficiency by implementing computations in memory, not in the processor, so that the data movements are eliminated. 3DICs is a promising solution to implement PIM, where the computation logic can be implemented on one layer and the memory array on another layer. In 2DIC, this is not cost-effective as computation logic needs more metal layers than memory arrays.

5.4.3 Lossy Compression

Lossy compression techniques remove unnecessary data. This is used for image, sound, and video compression since it can cause significant reduction in file size with no significant quality reduction [38]. Moreover, truncating the lower-bits in floating point data can be used.

5.5 Evaluation: Case Studies

5.5.1 Image Processing as a Case Study

An example from image processing domain that highlights the efficiency of the approximate computation is shown in Fig. 5.12, where a speedup factor of 2× can be obtained over the accurate solution with an acceptable accuracy or approximation error. The resource usage improvement and the accuracy are shown in Figs. 5.13 and 5.14, respectively.

5.5.2 CORDIC Algorithm as a Case Study

The CORDIC algorithm is a clever method for accurately computing trigonometric functions using only additions, bit shifts, and a small lookup table. The Algorithm is working as follows: If a vector V with coordinates (x, y) is rotated through an

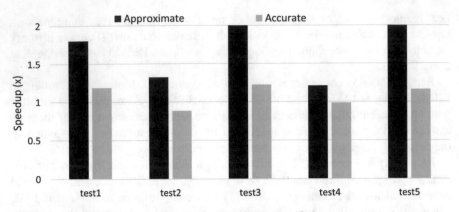

Fig. 5.12 Speedup factor between approximate and accurate solution

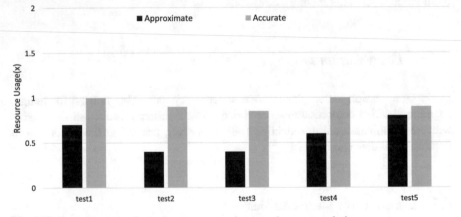

Fig. 5.13 Resource usage factor between approximate and accurate solution

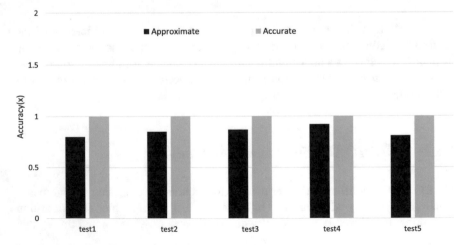

Fig. 5.14 Accuracy factor between approximate and accurate solution

angle Ø, then a new vector V′ with new coordinates (x′, y′) is formed, where x′ and y′ can be obtained using x, y, and Ø from the following method [39, 40].

$$x' = x\cos\emptyset - y\sin\emptyset, y' = x\sin\emptyset + y\cos\emptyset$$

By factoring out a {cos Ø} from both sides, resulting equation will be in terms of the tangent of the angle Ø. Next, if it is assumed that the angle Ø is being an aggregate of small angles α, and composite angles are chosen such that their tangents are all inverse powers of two, then this equation can be rewritten as an iterative formula:

$$x' = \cos\emptyset\left(x - y\tan\emptyset\right) \tag{5.3}$$

$$y' = \cos\emptyset\left(y + x\tan\emptyset\right) \tag{5.4}$$

$$z' = z + \emptyset \tag{5.5}$$

Here, Ø is the angle of rotation and z is the argument.

The multiplication by the tangent term can be avoided if the rotation angles and $\tan(\emptyset)$ are restricted so that $\tan(\emptyset) = 2^{-i}$. In digital hardware this denotes a simple shift operation.

Furthermore, if those rotations are performed iteratively and in both directions, every value of $tan(\emptyset)$ is representable. With $\emptyset = \tan^{-1}(2^{-i})$ the cosine term could also be simplified and since $\cos(\emptyset) = \cos(-\emptyset)$ it is constant for a fixed number of iterations. This iterative rotation can now be expressed as:

$$x = r\cos\theta, y = r\sin\theta \tag{5.6}$$

$$v' = \begin{pmatrix} x' \\ y' \end{pmatrix} = \begin{pmatrix} \cos\emptyset & -\sin\emptyset \\ \sin\emptyset & \cos\emptyset \end{pmatrix} \begin{pmatrix} x \\ y \end{pmatrix} \tag{5.7}$$

$$x_{i+1} = k_i \left[x_i - d_i y_i 2^{-i} \right] \tag{5.8}$$

$$y_{i+1} = k_i \left[y_i + d_i x_i 2^{-i} \right] \tag{5.9}$$

$$z_{i+1} = z_i - d_i \alpha_i \tag{5.10}$$

where, i denotes the number of rotations required to reach the required angle of the required vector $K_i = \cos(\tan^{-1}2^{-i})$ and $d = \pm 1$. α_i is the product of the K_i and represent the so-called K factor.

$$K = \prod_{i=0}^{N-1} k_i = \cos\emptyset_0 \cos\emptyset_1 \cos\emptyset_2 \ldots\ldots \cos\emptyset_{N-1} \cong 0.60725 \tag{5.11}$$

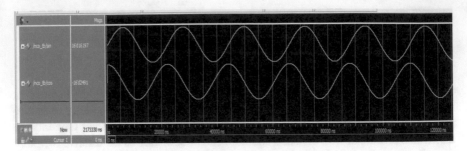

Fig. 5.15 Sine and Cosine plot. Frequency step = 2.8°

The CORDIC Algorithm can be used in iterative mode, to simplify each rotation, picking α_i (angle of rotation in ith iteration) such that $\alpha_i = (d_i\,0.2^i)$. d_i is such that it has value +1 or −1 depending upon the rotation, i.e., $d_i \in \{+1, -1\}$. Then:

$$x_{i+1} = \left[x_i - d_i y_i 2^{-i} \right] \tag{5.12}$$

$$y_{i+1} = \left[y_i + d_i x_i 2^{-i} \right] \tag{5.13}$$

$$z_{i+1} = z_i - d_i \tan^{-1} 2^{-i} \tag{5.14}$$

The computation of x_{i+1} or y_{i+1} requires an i-bit right shift and an add/subtract. If the function $\tan^{-1}(2^{-i})$ is pre computed and stored in table for different values of i, a single add/subtract suffices to compute $z_i + 1$. Each CORDIC iteration thus involves two shifts, a table lookup and three additions. Figure 5.15 shows sine and cosine plot for frequency step = 2.8°. The cordic algorithm is used in Bluetooth systems [47, 48].

5.5.3 HEVC Algorithm as a Case Study

HEVC similar to other video codec depends on splitting the frame into small blocks and address each block. Compared with the 8 × 8 pixels macro-blocks of the previous edition of AVC, HEVC uses coding tree units (CTUs) of 64 × 64 pixels. These CTUs can be further divided into smaller code units (CU) or code blocks (CB) down to 4 × 4 pixels. HEVC also enhances frames prediction by intra frame and inter frame prediction. Inter frame prediction—also called temporal prediction—depends on the repeated pixels between different frames where one frame is said to be the initial frame (I frame) and other frames having complete pixels and other predicted frames (P frames) are predicted from it and from other P frames. One of the biggest advances in HEVC is the spatial compression. The intra prediction has 35 direction of prediction, on DC, another is planner as well as 33 angular directions. As the angular prediction directions increases that make it applicable to obtain more code blocks from previously decoded blocks. The prediction units (PU) are usually smaller than the code unit unless the code unit is equal to 4×4, then both will have the same size. As the size of the code unit increases, the used prediction directions increase till all the directions are used when the CU is 64×64 that is not achievable using prior codecs. The difference between the original photo and the predicted one is called the residual. The residuals are the actual bits sent. These bits are quantized and transformed using discrete cosine transform or discrete sine transform. Entropy encoding Context-adaptive binary arithmetic coding (CABAC) is used to losslessly compress the residuals before sending them [41, 42]. HEVC is asymmetric codec which uses more computation in encoding stage than decoding stage. The architecture is quite similar except that in encoder we have motion estimators and intra picture estimators. Motion estimator creates motion vectors for different blocks to show how the blocks move between different frames. Prediction error accompanies the motion vector to fix any error in reconstructing the predicted image. On the other side, in the decoder the estimators are replaced by motion predictor and intra picture predictors. The function of the estimators is to estimate the changes in the pixels for next frames to decrease the encoding time. The predictors are used to predict the changed pixels and the ones that should be remained with the same value to complete the decoding process. For generating the prediction error, we have to reconstruct the image at the encoder side and monitor the differences between the original image and the predicted image to be sent to the decoder side. This implies that the encoder stage has both encoder and decoder processes, that is why it takes more time than decoding. The proposed architecture is shown in Fig. 5.2. The blocks are integrated to perform the encoding and decoding processes. Transform, Scaling, and quantization block is used to perform discrete cosine transformation (DCT) to remove the high frequency components in the image to reduce the power of the used in the prediction and transmitting the data. De-blocking and SAO filter are used to smoothing the image and removing boundaries between the CTUs. CABAC is used after that to generate the encoded bit-stream to be transmitted to the decoder side. Nowadays video delivery systems rely on video compression to send video data. Dedicated hardware platforms is considered as a better option for high video quality compression platforms as they provide fast performance as the hardware component do certain processes. Hardware platforms use ASIC (Application-Specific Integrated Circuits) and FPGA (Field Programmable Gate Arrays) for encoding processing. Software platforms use general purposes processors; this limits the compression process as processors perform sequential processes and limits the parallelism. Implementing the algorithm to a hardware platform increases the parallelism of the processes as well as decreasing delay. It will also improve power usage since the number of parallel operations increases; the frequency of each block may be reduced without affecting the throughput. However, reducing the frequency decreases the power consumption, allowing us to use the algorithm in low power applications as IoT and battery-based devices. Hence, our target is to implement the HEVC algorithm using hardware ASIC or FPGA platforms trying to reach a design with the optimum speed and power.

5.5.4 Software-Based Fault Tolerance Approximation

Although hardware-based fault mitigation techniques are very effective, they are based on hardware redundancy. Instead, there are techniques based on the use of redundant software, known as Software Implemented Hardware Fault Tolerance (SIHFT) techniques. These are suitable for processor-based systems, avoiding the modification of hardware components when designing fault-tolerant systems. Some of them are designed for fault detection, while others have correction capabilities [43]. More copies of the instructions in a program are executed using redundant registers, and majority voters are added before several critical instructions. Restrepo-Calle et al. [44] proposed a method that reduces overheads associated with software-based fault tolerance techniques, using the approximate computing paradigm. The proposal presented focuses mainly on reducing runtime overheads, relaxing the program before it is hardened using SIHFT approaches. This results on reduced computational costs, sacrificing some accuracy in the results.

5.6 Conclusions

Sacrificing exact calculations to improve performance in terms of run-time, power, and area is at the foundation of approximate computing. The survey shows that the approximate computing is a promising paradigm towards implementing ultra-low-power systems with an acceptable quality for applications that do not require exact results. Approximate computing exposes trade-offs between accuracy and resource usage. It is an important paradigm especially for resource-constrained embedded systems. Approximate computing is a promising technique to reduce energy consumption and silicon area. The approximate computing paradigm seeks to improve the efficiency of a computer system by reducing the quality of the results. AC aims at relaxing the bounds of exact computing to provide new opportunities for achieving gains in terms of energy, power, performance, and/or area efficiency at the cost of reduced output quality, typically within the tolerable range.

References

1. Q. Xu, T. Mytkowicz, N.S. Kim, Approximate computing: A survey. IEEE Design Test **33**, 8–22 (2016)
2. S. Mittal, A survey of techniques for approximate computing. ACM Comput. Surv. **48**, 62:1–62:33 (2016)
3. Kulkarni, et al., *Trading Accuracy for Power with an Under Designed Multiplier Architecture* (2011)
4. Z. Zhao, W. Qian, *A General Design of Stochastic Circuit and Its Synthesis* (Design, Automation, and Test in Europe (DATE), Grenoble, 2015)
5. V. K. Chippa, S. T. Chakradhar, K. Roy, A. Raghunathan, *Analysis and Characterization of Inherent Application Resilience for Approximate Computing*, In The 50th Annual Design Automation Conference 2013, DAC'13. ACM (2013), pp. 1–9
6. J. Han, M. Orshansky, *Approximate Computing: An Emerging Paradigm for Energy-Efficient Design*, In Proc. of the 18th IEEE European Test Symposium. IEEE (2013), pp. 1–6
7. H. Esmaeilzadeh, A. Sampson, L. Ceze, D. Burger, Neural acceleration for general-purpose approximate programs. Commun. ACM **58**(1), 105–115 (2015)
8. V. Gupta, D. Mohapatra, A. Raghunathan, K. Roy, Low-power digital signal processing using approximate adders. IEEE Trans Comp Aid Design Integr Cir Sys **32**(1), 124–137 (2013)
9. K. Nepal, Y. Li, R. I. Bahar, S. Reda, *ABACUS: A Technique for Automated Behavioral Synthesis of Approximate Computing Circuits*. In 2014 Design, Automation Test in Europe Conference Exhibition (DATE) (2014), pp. 1–6. https: //doi.org/10.7873/DATE.2014.374
10. Q. Xu et al., *Approximate Computing: A Survey* (IEEE Design & Test, 2016)
11. Q. Zhang, T. Wang, Y. Tian, et al., *ApproxANN: An Approximate Computing Framework for Artificial Neural Network*. In: Proceedings of the 2015 Design Automation & Test in Europe Conference & Exhibition. EDA Consortium (2015), pp. 701–706
12. D. Mohapatra, V.K. Chippa, A. Raghunathan, et al., *Design of Voltage-Scalable Meta-Functions for Approximate Computing*. In: Design, Automation &Test in Europe Conference & Exhibition (DATE), 2011. (IEEE, 2011), pp. 1–6
13. S. Venkataramani, V.K. Chippa, S.T. Chakradhar, et al. *Quality Programmable Vector Processors for Approximate Computing*. In: IEEE/ACM International Symposium on Microarchitecture (ACM, 2013), pp. 1–12
14. K. Salah, *Design and FPGA Implementation of Non-data Aided Timing and Carrier Recovery Techniques for EDR Bluetooth Standard*. Signal processing algorithms, architectures, arrangements, and applications (SPA) (IEEE, 2008)
15. V. Gupta, D. Mohapatra, A. Raghunathan, K. Roy, Low-power digital signal processing using approximate adders. IEEE Trans. On CAD of Integr. Circuits and Systems **32**(1), 124–137 (2013)
16. C.-H. Lin, I.-C. Lin, *High Accuracy Approximate Multiplier with Error Correction*. In: Computer Design (ICCD), 2013 IEEE 31st International Conference on. IEEE (2013), pp. 33–38
17. G.R. Morris, K. Abed, Mapping a Jacobi iterative solver onto a high performance heterogeneous computer. IEEE Trans Parallel and Distrib-Sys **24**(1), 85–91 (2013)
18. H. Fathalizadeh, *Solving Nonlinear Ordinary Differential Equations Using Neural Networks*" 2016 4th International Conference on Control, Instrumentation, and Automation (ICCIA) (2016), pp. 27–28
19. S. Keskar, Design and implementation of low-power digital signal processing using approximate adders. Int J Sci Res Eng Technol., ISSN 2278 – 0882 **4**(2) (2015)
20. W. J. Poppelbaum, C. Afuso, J. W. Esch, *Stochastic Computing Elements and Systems*, In Proc. Fall Joint Comput. Conf., Nov. 14–16 (1967), pp. 635–644. http://doi.acm.org/10.1145/1465611.1465696
21. H. Sim, S. Kenzhegulov, J. Lee, *DPS: Dynamic Precision Scaling for Stochastic Computing-based Deep Neural Networks*. In Proceedings of the 55th Annual Design Automation

Conference (DAC '18). ACM, New York, NY, USA (2018), Article 13, p. 6. https://doi.org/10.1145/3195970.3196028

22. H. Sim, J. Lee, *A New Stochastic Computing Multiplier with Application to Deep Convolutional Neural Networks*. In 2017 54th ACM/EDAC/IEEE Design Automation Conference (DAC) (2017), pp. 1–6. https://doi.org/10.1145/3061639.3062290

23. H. Sim, J. Lee, *Log-Quantized Stochastic Computing for Memory and Computation Efficient DNNs*. In Proceedings of the 24th Asia and South Pacific Design Automation Conference (ASPDAC '19). ACM (2019), pp. 280–285

24. W. El-Harouni, S. Rehman, B. S. Prabakaran, A. Kumar, R. Hafiz, M. Shafique, *Embracing Approximate Computing for Energy-Efficient Motion Estimation in High Efficiency Video Coding*. In 2017 Design, Automation & Test in Europe Conference & Exhibition (DATE). IEEE (2017), pp. 1384–1389

25. M. Franceschi, *Approximate FPGA Implementation of CORDIC for Tactile Data Processing using Speculative Adders*. IEEE NGCAS (2017)

26. M.A. Hanif, A. Marchisio, T. Arif, R. Hafiz, S. Rehman, M. Shafique, X-DNNs: systematic cross-layer approximations for energy-efficient deep neural networks. J Low Power Electr **14**(4), 520–534 (2018)

27. X. He, L. Ke, W. Lu, G. Yan, X. Zhang, *AxTrain: Hardware-Oriented Neural Network Training for Approximate Inference*. In Proceedings of the International Symposium on Low Power Electronics and Design (ISLPED '18). ACM, New York, NY, USA (2018), Article 20, p. 6. https://doi.org/10.1145/3218603.3218643

28. S. Mittal, A survey of techniques for approximate computing. ACM Comp Surv **48**(4), 62 (2016)

29. R. Soheil Hashemi I. Bahar, S. Reda, *DRUM: A Dynamic Range Unbiased Multiplier for Approximate Applications*. In IEEE/ACM Int. Conf. on Computer-Aided Design (ICCAD '15) (2015)

30. H. Jiang, C. Liu, N. Maheshwari, F. Lombardi, J. Han. *A Comparative Evaluation of Approximate Multipliers*. In IEEE/ACM Int. Symposium on Nanoscale Architectures (NANOARCH) (2016)

31. P. Coussy, *An Introduction to High-Level Synthesis*. In IEEE Design & Test of Computers (2009)

32. K. Salah, *Design and FPGA Implementation of Non-data Aided Timing and Carrier Recovery Techniques for EDR Bluetooth Standard*. Signal Processing Algorithms, Architectures, Arrangements, and Applications (SPA), 2008. (IEEE, 2008)

33. A. Aponte-Moreno, C. Pedraza, F. Restrepo-Calle, *Reducing Overheads in Software-based Fault Tolerant Systems using Approximate Computing*. IEEE (2019)

34. A. Momeni, J. Han, P. Montuschi, et al., Design and analysis of approximate compressors for multiplication. IEEE Trans. Comput. **64**(4), 984–994 (2015)

35. N. Sayed, F. Oboril, A. Shirvanian, R. Bishnoi, M. Tahoori, *Exploiting STT-MRAM for Approximate Computing*. 22nd IEEE European Test Symposium (ETS) (2017)

36. B. Zeinali, D. Karsinos, F. Moradi, *Progressive Scaled STT-RAM for Approximate Computing in Multimedia Applications*. IEEE Transactions on Circuits and Systems II: Express Briefs, (2017)

37. Kim et al., *Flipping Bits in Memory without Accessing Them: An Experimental Study of DRAM Disturbance Errors*. ISCA (2014)

38. C. M. Sadler, M. Martonosi, *Data Compression Algorithms for Energy-Constrained Devices in Delay Tolerant Networks*, Proceeding of ACM SenSys (2006)

39. K. Salah, *FPGA Implementation of Bluetooth 2.0 Transceiver*. In Proceedings of the 5th WSEAS international conference on System science and simulation in engineering (World Scientific and Engineering Academy and Society (WSEAS), 2006)

40. G.J. Sullivan et al., Overview of the high efficiency video coding (HEVC) standard. IEEE Trans. Circ Syst Video Technol **22**(12), 1649–1668 (2012)

41. Data compression, En.wikipedia.org (2018). https://en.wikipedia.org/wiki/Data_compression

42. M. Nourazar, V. Rashtchi, A. Azarpeyvand, F. Merrikh-Bayat, Code acceleration using Memristor-based approximate matrix multiplier: Application to convolutional neural networks. IEEE Trans Very Large Scale Integr Syst **26**, 12 (2018)
43. L. Renganarayana, V. Srinivasan, R. Nair, and D. Prener, *Programming with Relaxed Synchronization*. In Proceedings of the 2012 ACM Workshop on Relaxing Synchronization for Multicore and Manycore Scalability - RACES '12, New York: ACM Press, 2012, p. 41
44. F. Restrepo-Calle, A. Martinez-Alvarez, S. Cuenca-Asensi, A. Jimeno-Morenilla, Selective SWIFT-R. a flexible software-based technique for soft error mitigation in low-cost embedded systems. J. Electron. Test. **29**(6), 825–838 (2013)

Chapter 6
Near-Memory/In-Memory Computing: Pillars and Ladders

6.1 Introduction

Dynamic Random Access Memory (DRAM) is a vital component in nearly all computing systems such as servers, cloud platforms, and embedded devices. The need for fast analysis of data is increasing. Thus, DRAM is becoming a significant bottleneck due to the high energy and latency associated with data movement. The cost of data movement is a fundamental issue with the **processor-centric** paradigm of contemporary computer systems where CPU is considered the master in the system, and computation is performed only in the processor. To overcome these limitations on idea is to place computation mechanisms in or near where the data is stored. Smart self-powered embedded devices have become a part of daily life. These devices typically run **lightweight** applications like sensing to detect environmental changes. However, powering such devices is a critical challenge because of frequent power losses or outage as to the unpredictable nature of the energy source, the system will be subject to power outages. Due to the recent advancement in the nonvolatile memory (NVM) technologies, nonvolatile computing systems are now considered as a promising technology for self-powered devices and battery-less designs. These systems are able to sustain computations under unstable power, by quickly saving the state of the full system in a nonvolatile fashion [1]. NVPs implement complex hybrid memory elements (nvFF and nvSRAM) that allow for very fast parallel backup and **restore** operations. In this chapter, we explore different approaches to enable processing in and near memory [2–5].

© Springer Nature Switzerland AG 2020
K. S. Mohamed, *Neuromorphic Computing and Beyond*,
https://doi.org/10.1007/978-3-030-37224-8_6

6.2   Classical Computing: Processor-Centric Approach

Classical computing using multicores are shown in Fig. 6.1. L1 cache "level 1 cache" is SRAM memory that is very close to the CPU. The level-2 cache is a bit farther away on the chip. L2 cache is much larger, since more "real estate" is devoted to memory. The Intel Core i7 has 1 Mbyte cache. In modern multicore processors, the cores share the L3 cache, which is typically 8–12 Mbyte. Cache memory has two major problems; that is why we do not make all the storage in computer as a cache solution. It consumes huge amounts of power compared to DRAM memory (a flip-flop has about 16 transistors; a DRAM cell uses only one). This means if more cache were used, the cost of a computer would go up dramatically, due to the cost of extra power to run it, and cost of cooling the computer.

Processing options are shown in Fig. 6.2, where it can be CPU, GPU, or FPGA. Memory hierarchy is shown in Fig. 6.3. Memory is physically arranged so that fastest elements (registers) are closest to the CPU and slower elements are

Fig. 6.1 Classical computing: multicores

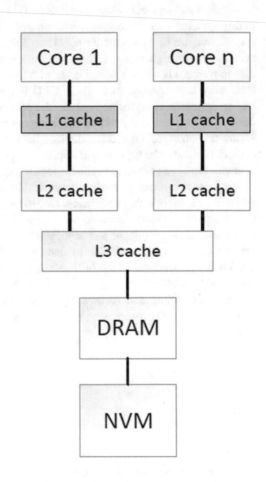

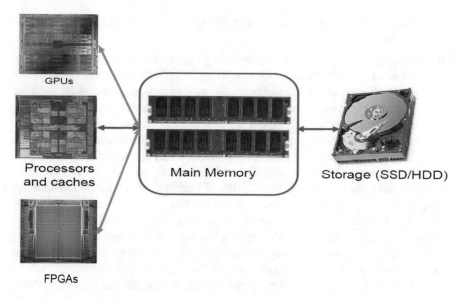

Fig. 6.2 Processing options

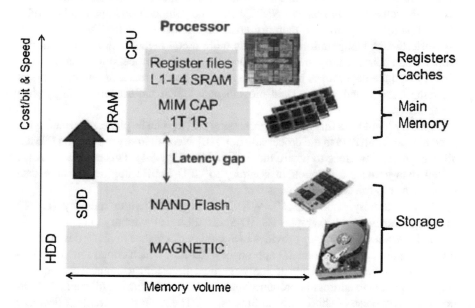

Fig. 6.3 Memory hierarchy

progressively farther away. So, we use a small amount of very fast SRAM cache memory which are physically near the computer, a substantial amount of DRAM, which is still very fast, as the main "working memory," and HDD or flash memory for large program storage.

6.3   Near-Memory Computing: Data-Centric Approach

Main memory is an important component in all computing systems, so it needs to be scaled in terms of efficiency to maintain performance growth and scaling pros. Recent technology applications and needs have led to new requirement for the main memory (more capacity, more bandwidth, less power consumption, and computations near memory as Data movement is much more energy-hungry than computation) [6, 7]. DRAM and flash memories do not satisfy all these requirements. There is a memory capacity gap with core count doubling every 2 years and DRAM capacity doubling every 3 years. Besides, DRAM consumes power even when not used. Moreover, according to ITRS, Scaling DRAM beyond 40-35 nm is challenging. Also, there is a problem with some recent DRAM devices in which repeatedly accessing a row of memory can cause bit flips in adjacent rows [8]. So, there is still a room for improvement in the current technology or development of emerging technologies such as new memory architectures (3D memory enables computations near memory), new memory technologies (NVM), or hybrid solutions [9].

Near-memory computing (NMC) provides high bandwidth and high energy efficiency by implementing computations in memory, not in the processor, so that the data movements are eliminated. Three-dimensional integrated circuits (3DICs) is a promising solution to implement NMC, where the computation logic can be implemented on one layer and the memory array on another layer. In 2DIC, this is not cost-effective as computation logic needs more metal layers than memory arrays. Both in- and near-memory are aimed at boosting the data processing functions. The aim is to reduce data moves between the memory and a processor which results in reducing latency and power. Data movement is much more energy-hungry than computation.

NMC try to address this issue by processing part of data in-place, eliminating the need to transfer all data to the processing unit [10]. Near-memory computing (NMC) aims at processing close to where the data resides (Fig. 6.4). This approach is also called **data-centric approach** in contrary to CPU-centric approach where data should move to the core for processing.

Hybrid memory cube (**HMC**), WIDE-IO, high bandwidth memory (**HBM**) memory controllers are examples for 3D-Stacked Logic+ Memory [11].

3D integration technology provides increased performance in many design criteria as compared to the current 2D approaches. 3D-ICs, which contain multiple layers of active devices, extensively utilize the vertical dimension to connect components and are expected to address interconnect delay related problems in planar (2D) technologies, by the use of short wires in 3D designs. These shorter wires will decrease the average load capacitance and resistance and decrease the number of repeaters which are needed to regenerate signals on long wires and to enable the integration of heterogeneous technologies. In the 3D design, an entire (2D) chip is divided into a number of different blocks, and each one is placed on a separate layer of silicon that are stacked on top of each other. This may be exploited to build SoC by placing different circuits with performance requirements in different layers (Fig. 6.5).

Fig. 6.4 Near-memory computing (NMC). For example, HMC-assisted processing. The HMC has multiple DRAM dies stacked on top of a logic layer that can provide the ability of computation with high memory access parallelism

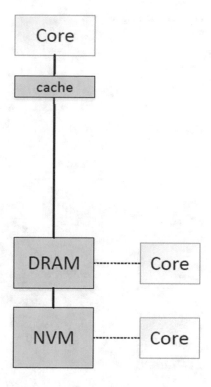

6.3.1 HMC

An illustrative example of HMC is shown in Fig. 6.6 [12, 13]. HMC uses 3D single packaging of 4 or 8 DRAM memory dies and one logic die collected together using through-silicon vias (TSV) and micro-bumps with smaller physical footprints. HMC exponentially is more power efficiency and energy savings, utilizing 70% less energy per bit than DDR3 DRAM technology. A single HMC can provide more than 15× the performance of DDR3 module, which increase bandwidth. HMC reduced latency with lower queue delays and higher bank availability. It can keep up with the advancements of CPUs and GPUs. HMC uses standard DRAM cells but its interface is incompatible with current DDR2 or DDR3 implementations. It has more data banks than classic DRAM of the same size. HMC memory controller is integrated into memory package as a separate logic die. The logic base manages multiple functions for HMC, like all HMC I/O, mode and configuration registers and data routing and buffering between I/O links and vault. A crossbar switch is an implementation example to connect the vaults with I/O links. The external I/O links consist of multiple serialized 4 or 8 links. Each link with a default of 16 input lanes and 16 output lanes for full width configuration, or 8 input lanes and 8 output lanes for half width configuration [1].

Fig. 6.5 3D integration example

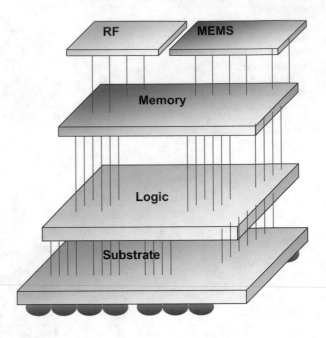

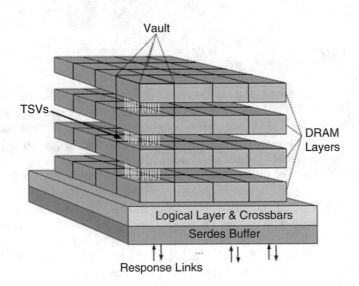

Fig. 6.6 HMC example

6.3.2 WideIO

WideIO mobile DRAM uses chip-level dimensional (3D) stacking with through-silicon via (TSV) interconnects and memory chips directly stacked upon a system on a chip (SOC). WideIO DRAM major advantage over its predecessors (such as LPDDR DRAM) is that, it offers more bandwidth at lower power. WideIO is the first interface standard for 3D die stacks and offering a compelling bandwidth and power benefit. WideIO is particularly suited for applications requiring increased memory bandwidth up to 17GBps, such as 3D Gaming and HD video. WideIO will provide the ultimate in performance, energy efficiency and small size for smart phones, tablets, handheld gaming consoles and other high-performance mobile devices. Given the ever-growing hunger for memory bandwidth and the need to reduce memory power in many applications, WideIO is the first standard for stackable WideIO DRAMs. This standard widens the conventional 32 bit DRAM interface to 512 bits [14].

6.3.3 HBM

HBM (High Bandwidth Memory) is a new type of DRAM-based memory chip with low power consumption, ultra-wide communication lanes, and a revolutionary new stacked configuration. HBM uses 128-bit wide channels. It can stack up to eight of them for a 1024-bit interface. The total bandwidth ranges from 128GB/s to 256GB/s. Each memory controller is independently timed and controlled. Future GPUs built with HBM might reach 1 TB/s of main memory bandwidth. HBM is designed for high-performance GPU environments as it is cheaper than HMC [15]. An example is shown in Fig. 6.7.

6.4 In-Memory Computing: Data-Centric Approach

Computation in-memory uses novel devices such as memristors, phase change memory (PCM), spin-transfer torque RAM (STT-RAM), and resistive random access memory (ReRAM), where they have inherent computation capability (Fig. 6.8).

6.4.1 Memristor-Based PIM

The logic state of memristors depends upon the resistance of the device, which is controlled by the charge through it. They have fast switching speed, low switching energy, and high scalability, making them suitable for dense and fast PIM solutions.

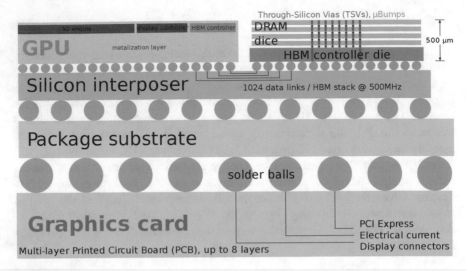

Fig. 6.7 HBM example

Some compute logic at the periphery of the memory by modifying the memory sense amplifiers. They read the stored data from the memory, use transistor-based circuits to process data, and store the results back to the memory [16].

Memristor is a two terminal component where its resistance mainly lies on the magnitude, applied voltage, and polarity. As the voltage is not applied, then the resistance leftover, and this makes this as a nonlinear and memory component. The memristor uses a titanium dioxide (TiO_2) like a resistive material. It works superior to other kinds of materials like silicon dioxide. When the voltage is given across the platinum electrodes, then the Tio_2 atoms will spread right or left in the material based on voltage polarity which makes thinner or thicker, therefore giving a transform in resistance (Fig. 6.9)..

6.4.2 PCM-Based PIM

Phase-change memory (PCM) is a nanoscale memory device based on the property of certain compounds of Ge, Te, and Sb that exhibit drastically different electrical characteristics depending on their atomic arrangement. In the disordered amorphous phase, these materials have very high resistivity, while in the ordered crystalline phase, they have very low resistivity. A PCM device consists of a nano-metric volume of this phase change material sandwiched between two electrodes as shown in Fig. 6.10 [17].

It is possible to perform in-place computation with data stored in PCM devices. The essential idea is not to treat memory as a passive storage entity, but to exploit the physical attributes of the memory devices, and thus realize computation exactly

Fig. 6.8 In memory computing: uses novel devices such as memristors and phase change memory (PCM), where they have inherent computation capability. Processing-in-memory (PIM) has been explored as a promising solution to providing high bandwidth

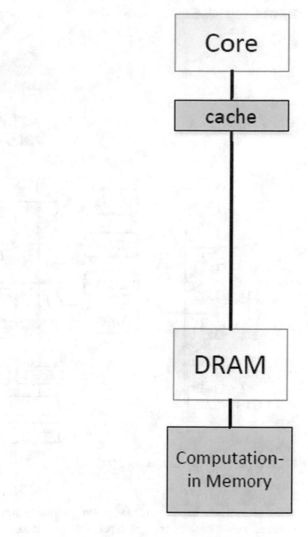

Fig. 6.9 Memristor structure

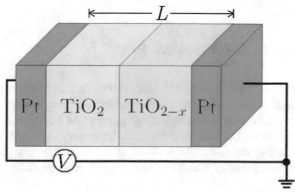

Fig. 6.10 PCM structure

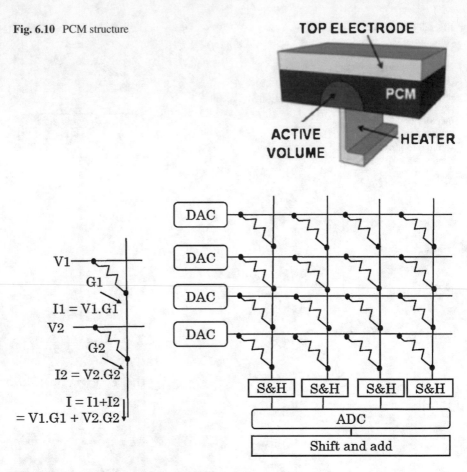

Fig. 6.11 Matrix-vector multiplication [37]

at the place where the data are stored. Programming strategy based on uniform short pulse sequence enables gradual depression (nonstationary regime).

6.4.3   ReRAM-Based PIM

Resistive RAM (ReRAM) offer distinct advantages over conventional CMOS (complementary metal–oxide–semiconductor)-based designs. ReRAM is a nonvolatile memory with near-zero leakage energy and high density. The ReRAM state reflects the current passed through it in the history. ReRAM supports operations such as analog matrix-vector multiplication (Fig. 6.11), search and bitwise operations within memory which facilities energy-efficient designs.

6.4.4   STT-RAM-Based PIM

Spin-transfer torque RAM (STT-RAM) has higher latency and writes energy as compared to SRAM and DRAM. STT-MRAM utilizes bi-stable tunnel magneto-resistance in a magnetic tunnel junction (MTJ) for data storage [18]. It consists of two "ferromagnetic layers" separated by a thin metallic oxide tunneling layer. The relative angular momentum or spin of the two "ferromagnetic layers" is leveraged to store binary data. The layers can be in two possible orientations: one where both layers have the same or parallel spins and the other where both layers have opposing or antiparallel spins [19].

An STT-RAM bit-cell consists of an access transistor and magnetic tunnel junction (MTJ), as shown in Fig. 6.12. An MTJ in turn consists of a pinned layer that has a fixed magnetic orientation and a free layer whose magnetic orientation can be switched. The magnetic layers are separated by a tunneling oxide. The relative magnetic orientation of the free and pinned layers determines the resistance offered by the MTJ [20]. It targets NVMe applications such as SSDs.

6.4.5   FeRAM-Based PIM

Ferroelectric RAM (FeRAM, F-RAM or FRAM) is a random-access memory similar in construction to DRAM but using a ferroelectric layer instead of a dielectric layer to achieve nonvolatility. FeRAM is one of a growing number of alternative nonvolatile random-access memory technologies that offer the same functionality as flash memory. FeRAM's advantages over flash include lower power usage, faster write performance, and a much greater maximum read/write endurance.

Fig. 6.12 STT-RAM structure

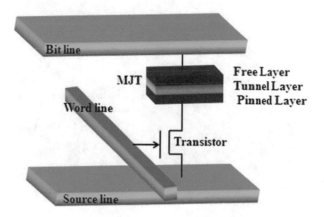

6.4.6 NRAM-Based PIM

Carbon nanotube NRAM has DRAM-like speed, nonvolatility, and low cost. Nanotubes can switch state in picoseconds, so it is very fast. NRAM is a memory cell made up of an interlocking matrix of CNTs, either touching or slightly separated, leading to low or higher resistance states, respectively. The small movement of atoms, as opposed to moving electrons for traditional silicon-based memories, renders NRAM with a more robust endurance and high temperature retention. NRAMs are expected to be a disruptive replacement for DRAM SRAM, and NAND flash memories [21].

Carbon nanotube (CNT) is cylinders of graphite, same as the material in a pencil. CNT can be metallic or semiconductor, it all depends on the angle of rolling and diameter of the nanotube. CNT-based TSV has excellent current-carrying capability higher than copper, as it does not exhibit skin effect at high frequency. Moreover, it has better thermal stability than copper as it works fine at higher temperature, thus eliminating the problem of electromigration with Cu interconnects. CNT is a fairly new molecular structure of carbon and has been applied to semiconductor industry recently [22]. It has unique electrical and mechanical properties that it can be shaped to act as conductor, semiconductor, and insulator. CNT is able to reduce semiconductor manufacturing processing steps that incorporate copper, such as the forming of insulator to prevent copper from diffusing through the gate. CNTs are cylindrical structures based on the hexagonal lattice of carbon atoms that covalently bond together and forms crystalline graphite. CNTs can be looked at as single molecules, due to their small size. There are two major types of CNTs: single-walled carbon nanotubes (SWNT) and multi-walled carbon nanotubes (MWNT). Multi-walled carbon nanotubes can be described as an assembly of concentric single-walled carbon nanotubes with different diameters (Fig. 6.13).

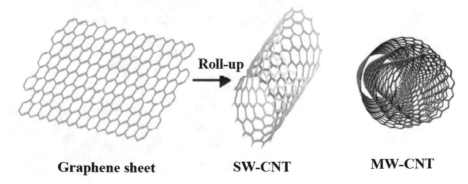

Graphene sheet   **SW-CNT**   **MW-CNT**

Fig. 6.13 CNT structure

Table 6.1 Comparison between different new memories [38]

Technology	FeRAM	MRAM	ReRAM	PCM	DRAM	NAND Flash
Nonvolatile	Yes	Yes	Yes	Yes	No	Yes
Endurance	10^{12}	10^{12}	10^6	10^8	10^{15}	10^3
Write Time	100ns	~10ns	~50ns	~75ns	10ns	10μs
Read Time	70ns	10ns	10ns	20ns	10ns	25μs
Power Consumption	Low	Medium/Low	Low	Medium	Very High	Very High
Cell Size (f^2)	15-20	6-12	6-12	1-4	6-10	4
Cost ($/Gb)	$10/Gb	$30-70/Gb	Currently High	$0.16/Gb	$0.6/Gb	$0.03/Gb

6.4.7 Comparison Between Different New Memories

Comparison between different new memories in terms of endurance, write/read time, and power consumption are shown in Table 6.1. Figure 6.14 shows a summary of different memory technologies.

6.5 Techniques to Enhance DRAM Memory Controllers

Memory is an important component in all computing systems. So, it needs to be scaled in terms of efficiency to maintain performance growth and scaling pros. Recent technology applications and needs have led to new requirement for memory (more capacity, more bandwidth, less power consumption, and computations near memory as data movement is much more energy-hungry than computation).

DRAMs do not satisfy all these requirements. There is a memory capacity gap with core count doubling every 2 years and DRAM capacity doubling every 3 years. Moreover, DRAM consumes power even when not used [23–28].

Therefore, there is still a room for improvement in the current technology or development of emerging technologies. DRAM improvements are shown in Fig. 6.15.

A DRAM is a three-dimensional array of memory cells arranged as banks. A DRAM cell consists of a transistor and a capacitor, and their size has a direct impact on DRAM density. Cells in each bank are organized in rows and columns. A DRAM rank is a group of banks. For multichannel DRAMs, each channel has its own buses and consists of one or more ranks [29, 30]. Mainly, there are six families of DRAM memory interfaces or controllers called DDRx, LPDDRx, GDDRx, HBM, HMC, and WIDE-IO [31–36].

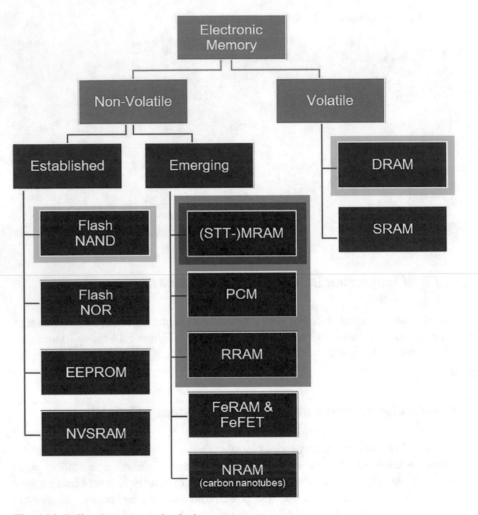

Fig. 6.14 Different memory technologies

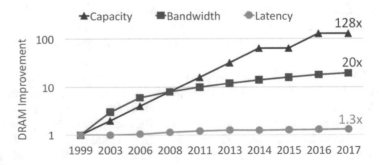

Fig. 6.15 DRAM improvements over time

Table 6.2 Dram memory interfaces trends and generations: performance comparison

	DDRx	LPDDRx	GDDRx	HBM	HMC	WIDE-IO
Recent specs	DDR5	LPDDR5	GDDR6	HBM	HMC2	WIDE-IO2
Power	+	+++	++	++++	++++++	+++++
Bandwidth/pin	+	++	+++	++++	++++++	+++++
Application	Servers	Smartphones	Graphics	Graphics	Servers	Smartphones
Capacity (GB)	8	8	8	8	8	8
Cost	+	++	+++	++++	+++++	++++++
Interface	Serial	Serial	Serial	Parallel	Serial	Parallel
JEDEC standard	Y	Y	Y	Y	N	Y
Complexity	++	+++	+++	++++++	++++	+++++
Technology	2D	2D	2D	3D	3D	3D

A comparison between them is shown in Table 6.2. Smartphones include fast memory devices based on LPDDRx standards, servers are based on DDRx, and graphic devices are based on GDRRx standards.

Memory Controllers support specific requests of the host and accounts for the constraints provided by the memory device. The host can be server, mobile processor, or GPU/FPGA. DDRx, LPDDRx, and GDDRx are 2D technology while HBM, HMC, and WIDE-IO are 3D technology.

3D technologies open a new era of memory architecture exploration. Compared to existing memory interface, TSV-based 3D technology provides better Bandwidth and less power consumption. Lower power consumption is achieved by lower capacitance of TSV [11]. 3D-stacked memories can be implemented in several architectures. One possible architecture is simply using TSVs to connect DRAM layers to the processor layer. Such memory architecture reduces the long memory access latency, but does not provide much bandwidth benefits as the individual memory structures in each layer are traditional 2D memory.

Another possible architecture is to stack individual storage-cell arrays in a 3D layer. TSVs are used to connect these memory arrays from different layer to logic layer.

In this chapter, we present a survey on the recent technologies and methods used in DRAM interfaces to improve performance in terms of BW, latency, capacity, area, and power.

6.5.1 Techniques to Overcome the DRAM-Wall

6.5.1.1 Low-Power Techniques in DRAM Interfaces

Power consumption is a function of operating frequency, operating voltage, architecture$_{software}$, architecture$_{hardware}$, and throughput of transmitter and receiver packets. The techniques used to improve power consumption in DRAM memory controllers are listed below:

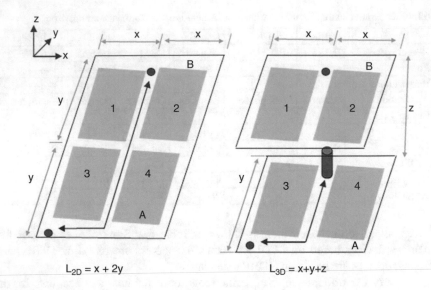

Fig. 6.16 Reduction in wire length where the original 2D circuit is implemented in two planes ($z < y$)

- **Dual data rate**: commands are sampled on positive and negative edges. So, the total number of clocks needed to send the commands are reduced.
- **Data bus inversion:** it is used for writes and read, as if the total count of "1" data bits is equal to or greater than five, then the data is inverted. This reduces the power consumption.
- **Error correction code**: as it is not needed to resend the erroneous data. This saves power.
- **Deep sleep mode:** which is an enhanced power saving mode.
- **Refresh management**: to reduce refresh time: Self-refresh or partial array self-refresh.
- **Write zero command**: which indicates that data is all 0's where some applications have up to 40% of writes as zeros. So, no data is sent on data bus which reduces power.
- **3D Packaging**: which results in Smaller form factor and less parasitics (Fig. 6.16).
- **Processing near memory**: perform computations near memory as data movement is much more energy-hungry than computation.
- **Shorter data path architecture.**
- **Data Copy mechanism**: whenever any data pattern is repeated over data bus, only the reference data is transferred, then the device recovers the original data by copying the reference data to other memory places.
- **Differential clocking:** which improve noise immunity and thus reducing error probability and FEC utilization.

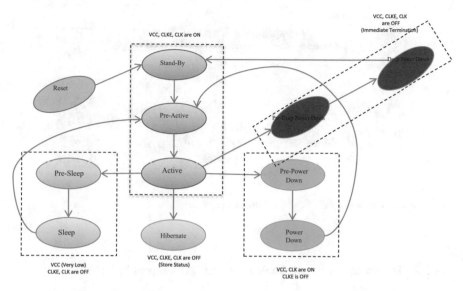

Fig. 6.17 Different power levels management

- **Voltage management**: reduce power supply or dynamic voltage and frequency scaling (DVFSC). Figure 6.17 shows different power levels management.
- **Low-leakage sense amplifier.**

6.5.1.2   High-Bandwidth and Low Latency Techniques in DRAM Interfaces

For DRAM, latency is important since random access is more frequent. Latency is the time consumed from the address or command input until the data gains access to the output latches. Bandwidth is the rate of data output per given unit of time. The techniques used to improve latency and bandwidth are listed below:

- **Dual bank architecture**: as it provides higher bandwidth efficiency [9]. Figure 6.18 shows an example for single-channel DRAM versus dual-channel DRAM.
- **Increase pin counts**: results in increasing the bandwidth.
- **3D Packaging**: as the changes in the DRAM interface when moving from 2D to 3D domain results in no need to have a single bidirectional bus as write and read data are routed on separated buses, which improve memory communication.
- **Smart refresh or all bank refresh**: because DRAM has tens of thousands of rows, so refreshing all of them in bulk incurs high latency. Instead, memory controllers send a number of refresh commands that are distributed throughout the retention time to trigger refresh operations.

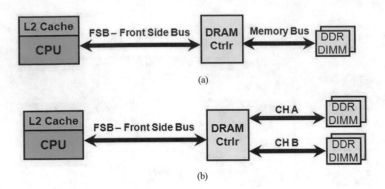

Fig. 6.18 (**a**) Single-channel DRAM, (**b**) dual-channel DRAM

6.5.1.3   High-Capacity and Small Footprint Techniques in DRAM Interfaces

The techniques used to increase capacity and reduce footprint are listed below:

- **Using low area materials for routing** [26].
- **Reducing the size of sense amplifier**.
- **3D Packaging**: it results in small footprint.
- **Transistor Scaling**.

6.6   Conclusions

PIM promises to solve the bandwidth bottleneck issues in today's systems. PIM enables moving from big data era to fast data era. PIM-enabled architectures achieve impressive performance results by integrating processing units into the memory. NV memories chips could lead the way to fast, energy-efficient hardware accelerators. PIM reduces data movement by performing Processing in Memory. DRAM memory controllers have gained a lot of popularity in recent years. They are widely used in applications ranging from smart phones to high-performance computers. These applications requires large amount of memory accessing. However, memory-wall is still a bottleneck. In this chapter, we introduce the most recent techniques used to enhance DRAM memory controllers in terms of power, capacity, latency, bandwidth, and area. The traditional information system adopts the architecture of separation of computing and storage. The data is transmitted between the CPU and the memory, and the power consumption is large and the speed is slow. The PIM adopts the architecture of storage and computing together, which eliminates the data transmission process and greatly improves the information processing efficiency.

References

1. Hybrid Memory Cube. *Technical Report Revision 1.0*, HMC (2013), www.hybridmemorycube.org
2. S. Ghose et al., *Enabling the Adoption of Processing-in-Memory: Challenges, Mechanisms, Future Research Directions.* arXiv:1802.00320 [cs:AR] (2018)
3. S. Ghose et al., *The Processing-in-Memory Paradigm: Mechanisms t Enable Adoption* (Beyond-CMOS Technologies for Next Generation Computer Design, 2019)
4. Z. Liu et al., *Concurrent Data Structures for Near-Memory Computing* (SPAA, 2017)
5. G.H. Loh et al., *A Processing in Memory Taxonomy and a Case for Studying Fixed-Function PIM* (WoNDP, 2013)
6. H. Asghari-Moghaddam, A. Farmahini-Farahani, K. Morrow, J.H. Ahn, N.S. Kim, Near-DRAM acceleration with single-ISA heterogeneous processing in standard memory modules. IEEE Micro **36**, 24–34 (2016)
7. M. Gao, G. Ayers, C. Kozyrakis, *Practical Near-Data Processing for In-Memory Analytics Frameworks*, in ACM International Conference on Parallel Architecture and Compilation (PACT) (2015)
8. Kim et al, *Flipping Bits in Memory Without Accessing Them: An Experimental Study of DRAM Disturbance Errors*, (ISCA, 2014)
9. S. Ghose, K. Hsieh, A. Boroumand, R. Ausavarungnirun, O. Mutlu, Enabling the adoption of processing-in-memory: Challenges, mechanisms, future research directions. arXiv preprint arXiv **1802**, 00320 (2018)
10. J. Ahn, S. Hong, S. Yoo, O. Mutlu, K. Choi, *A Scalable Processing In-Memory Accelerator for Parallel Graph Processing*, In Proc. ISCA, Portland, OR, USA (2015), pp. 105–117
11. O. Junior, et al., *A Generic Processing in Memory Cycle Accurate Simulator Under Hybrid Memory Cube Architecture* (2017)
12. J. Zhang, J. Li, *Degree-Aware Hybrid Graph Traversal on FPGA-HMC Platform*. In Proc. ACM/SIGDA Int. Symp. Field-Programmable Gate Arrays (2018), pp. 229–238
13. C.Y. Gui, L. Zheng, B.S. He, A survey on graph processing accelerators: challenges and opportunities. J. Comput. Sci. Technol. (2019)
14. Wide I/O Single Data Rate, *Technical Report Revision 1.0*, WideIO (2011)
15. C. Kim, H.-W. Lee, J. Song, *High-Bandwidth Memory Interface* (Springer, Berlin, 2014)
16. M. Imani et al., *Mpim: Multi-Purpose In-Memory Processing Using Configurable Resistive Memory*, In IEEE ASP-DAC (IEEE, 2017), pp. 757–763
17. A. Sebastian, M. Le Gallo, G.W. Burr, S. Kim, M. BrightSky, E. Eleftheriou, Tutorial: Brain-inspired computing using phase-change memory devices. J. Appl. Phys. **124**, 111101 (2018)
18. A.F. Vincent et al., Spin-transfer torque magnetic memory as a stochastic memristive synapse for neuromorphic systems. IEEE Trans Biomed Circ Syst **9**, 166–174 (2015)
19. S. Peng, Y. Zhang, M. Wang, Y. Zhang, W. Zhao, Magnetic tunnel junctions for spintronics: principles and applications, in *Wiley Encyclopedia of Electrical and Electronics Engineering*, ed. by J. Webster, (Wiley, New York, 2014), pp. 1–16
20. S. Chatterjee, M. Rasquinha, S. Yalamanchili, S. Mukhopadhyay, A scalable design methodology for energy minimization of STTRAM: A circuit and architecture perspective. IEEE Trans Very Large Scale Integr Syst **19**(5), 809–817 (2011)
21. https://iopscience.iop.org/article/10.1088/1361-6528/aaaacb/pdf
22. K. Salah, *Characterization of SWCNT-Based TSV*. In: 16th International Power Electronics and Motion Control Conference and Exposition, Antalya, Turkey 21–24 Sept 2014
23. A. Kim, *Flipping Bits in Memory without Accessing Them: An Experimental Study of DRAM Disturbance Errors* (ISCA, 2014)
24. C. Weis, N. Wehn, L. Igor, L. Benini, *Design Space Exploration for 3D-Stacked DRAMs* (DATE, 2011)
25. Y.U. Lin, Sh. Peng, W. Hwang, *WIDE-I/O 3D-Staked DRAM Controller for Near-Data Processing System* (IEEE, 2017)

26. K. T. Malladi, U. Kang, M. Awasthi, H. Zheng, *DRAMScale: Mechanisms to Increase DRAM Capacity* (MEMSYS, 2016)
27. N. Chidambaram, *GemDroid: A Framework to Evaluate Mobile Platforms* (SIGMETRICS, 2014)
28. M. Hassan, H. Patel, *MCXplore: An Automated Framework for Validating Memory Controller Designs*. (Design, Automation & Test in Europe Conference & Exhibition (DATE), 2016)
29. B. Akesson, P. Huang, F. Clermidy, D. Dutoit, *Memory Controllers for High-Performance and Real-Time MPSoCs*. In: Proceedings of the seventh IEEE/ACM/IFIP international conference on hardware/software codesign and system synthesis (2011)
30. C. Kim, *High-Bandwidth Memory Interface* (Springer, Berlin, 2014)
31. DDR5 SDRAM Standard, JEDEC Standard
32. LPDDR5 SDRAM Standard, JEDEC Standard
33. Graphics Double Data Rate (GDDR6) SGRAM Standard, JEDEC Standard
34. High Bandwidth Memory (HBM) DRAM, JEDEC Standard
35. About Hybrid Memory Cube, Hybrid Memory Cube Consortium. http://hybridmemorycube.org/technology.html
36. WIDE I/O Technical Report Revision 2.0, JEDEC Standard
37. S. Mittal, *A Survey of ReRAM-Based Architectures for Processing-In-Memory and Neural Networks* (MDPI, 2018)
38. https://www.snia.org/sites/default/files/PMSummit/2018/presentations/14_PM_Summit_18_Analysts_Session_Oros_Final_Post_UPDATED_R2.pdf

Chapter 7
Quantum Computing and DNA Computing: Beyond Conventional Approaches

7.1 Introduction: Beyond CMOS

The limitations of Si-Based ICs are now causing the industry to identify at least three main research domains, called: (1) "More Moore," (2) "More than Moore," and (3) "Beyond CMOS" as depicted in Table 7.1. The "More Moore" domain is traditionally dealing with technologies related to the silicon-based CMOS. The "More than Moore" domain encompasses the engineering of complex systems that can combine, by heterogeneous integration techniques (in SoC or SIP), various technologies (not exclusively electronic) in order meet certain needs and challenging specifications of advanced applications. The "Beyond CMOS" domain deals with new technologies and device principles (i.e., from charge-based to non-charge-based devices, from semiconductor to molecular technology). This chapter introduces ongoing trends to work around Moore's law limitations: Beyond CMOS solutions [1].

7.2 Quantum Computing

In conventional computing, numbers are represented in binary format. Mathematical operations are done based on this binary representation. Due to Moore's law saturations, many alternatives rose to speed-up computations and quantum computing is one of these alternatives. Quantum Physics violate classical laws at a small scale. Qubit is the quantum binary representation.

Quantum computers are not limited to two states like today's computers. They encode information as quantum bits, or qubits, which can exist in superposition [2]. Superposition-quantum computers can represent both 0 and 1 as well as everything in between at the same time. Qubits can be carried as atoms, ions, photons or electrons and their respective control devices that are working together to act as computer

© Springer Nature Switzerland AG 2020
K. S. Mohamed, *Neuromorphic Computing and Beyond*,
https://doi.org/10.1007/978-3-030-37224-8_7

Table 7.1 Comparison between different trends levels from performance gains point of view [27]

Trends			Performance gains					
			Area	Speed throughput delay	Power	Noise	Thermal	Yield reliability
Technology level	More Moore	Scaling	✓	✓	✓			
		New architectures						
		SOI (silicon on insulator)	✓					
		Twin-well			✓			
		FinFET		✓				
		New materials						
		High-K		✓				
		Metal-G		✓				
		Strained-Si		✓				
	More than Moore	**New interconnects schemes**						
		NoC (network on chip)		✓	✓			
		3D (three dimensional)		✓	✓			✗
		Optical-interconnects		✓	✓		✗	
		Wireless interconnects	✓			✗		
	Beyond CMOS	**New devices**						
		Molecular computer CNT (carbon nanotubes) (nanowires) (Quantum-Dot)	✓	✓				
		Biological computer (DNA-computing)		✓				✗
		Quantum computer (SET: single electron transistor) (spin device)	✓					
Architectural level		Me (multicore)		✓	✓			
		DM (distributed memory)		✓				

(continued)

Table 7.1 (continued)

		Performance gains					
Trends		Area	Speed throughput delay	Power	Noise	Thermal	Yield reliability
Circuit/logic level	Adiabatic			✓			
	MTCMOS			✓			
	Multiple-Vdd			✓			
	Clock-gating			✓			
	Power-gating			✓			
	Asynchronous			✓			
	Pipelining			✓			
	Data-encoding			✓			
	Repeater	✓					
Software/OS level	Concurrency			✓			
	Partitioning			✓			
	Sleep mode			✓			

memory and a processor. Basically, a quantum computer can work on a million computations at once, while your desktop PC works on one [3–7]. For quantum computers, the bigger the problem, the better. Let us take Grover's algorithm as an example. This quantum algorithm is able to find a specific name among 100 million names, while using only 10,000 operations. If the same task was given to a classical computer, it would use 50 million operations. Furthermore, the process, which would take 31 years to be completed on a classical computer, could be done in only 9 h on a quantum computer. It has been predicted that quantum computers will reach the computational power and maturity sufficient to break existing public-key cryptography algorithms by 2031, or earlier [8]. Quantum computers are not intended to replace classical computers, they are expected to be a different tool we will use to solve complex problems that are beyond the capabilities of a classical computer, i.e., to speed up solutions to hard problems [4, 9–11].

7.2.1   Quantum Computing: History

Quantum computing studies the computers and universal computational models which make use of quantum mechanics phenomena. It is about the implementation of algorithms that sustain their work in quantum properties. The theory of quantum mechanics has some aspects that are very different from their classical counterpart, and sometimes there is no classical counterpart at all, as in the cases of quantum entanglement and superposition. Table 7.2 summarizes main quantum computing milestones.

Table 7.2 Quantum computing history

1982	Feynman proposed the idea of creating machines based on the laws of quantum mechanics instead of the laws of classical physics.
1985	David Deutsch developed the quantum **turing** machine, showing that quantum circuits are universal.
1994	Peter Shor came up with a quantum algorithm to factor very large numbers in very short time.
1997	Lov Grover develops a quantum search algorithm which takes only square root of time to search than classical computer.
1998	2 Qubit quantum computer.
2006	12 Qubit quantum computer.
2007	Google's D-Wave with 28 Qubits.
2017	D-Wave With 2000 Qubits.

7.2.2 Quantum Computing: What?

It uses of quantum phenomenon to perform computational operations. Operations are done at an atomic level. What can be done with single **qubits** is very limited. For more ambitious information processing and communication protocols, multiple bits must be used and manipulated. A qubit is essentially an atom showing quantum-mechanical behavior. Just as a regular bit, qubits are also used to represent 1/0 values, usually denominated by the up-spin or down-spin of the atom. Spin: An integral quality of all elemental particles and related to orbital angular momentum [12, 13].

7.2.3 Quantum Computing: Why?

- Parallel computations: Increase efficiency for iterative repetitive tasks.
- Cryptography: Cracking RSA encryption, use quantum entanglement to send the decode key for classical cryptography.
- Sorting: huge database management.
- Quantum protocols that send exponentially less bits than classical.
- Speed-up prime Factoring.

7.2.4 Quantum Computing: How?

Physical realization of quantum computing is done by quantum dot (Single Electron Transistor SET), where we use the Electron spin $|\uparrow>$, $|\downarrow>$. A one is spin up $|\uparrow>$. A zero is spin down $|\downarrow>$, as shown in Fig. 7.1.

Fig. 7.1 Quantum
computing: Spin

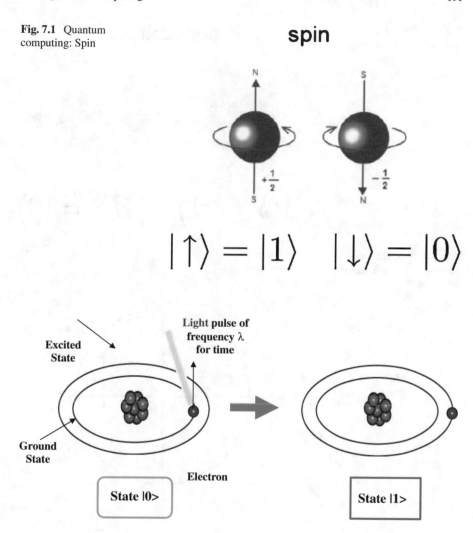

Fig. 7.2 Quantum computing: atomic level. It is about optical computing

A physical implementation of a qubit could also use the two energy levels of an atom. An excited state represents |1> and a ground state represents |0>. Figure 7.2 shows a physical explanation for quantum computing physical realization using the two energy levels of an atom.

For photons, it is about the polarization as depicted in Fig. 7.3.

Fig. 7.3 Quantum
computing: Polarization
for photons

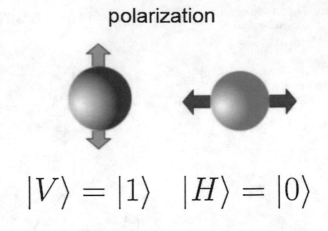

Fig. 7.4 A qubit may be
visualized as a unit vector
on the plane

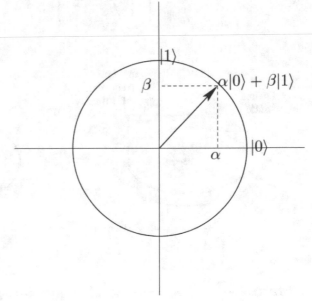

7.3   Quantum Principles

7.3.1   Bits Versus Qbits

A building block of classical computational devices is a two-state deterministic
system (0 or 1). A qubit may be visualized as a unit vector on the plane as shown in
Fig. 7.4. Qbit is described by [14, 15]:

$$\#\psi >= \alpha \ |0> + \beta \ |1> \tag{7.1}$$

where α and β are complex numbers called amplitudes and satisfying

$$|\alpha|^2 + |\beta|^2 = 1 \qquad (7.2)$$

For more qbits ψ is represented as

$$\#\psi \geq = c_{00}|00> + c_{01}|01> + c_{10}|10> + c_{11}|11> \qquad (7.3)$$

7.3.2   Quantum Uncertainty

The position and velocity of a particle are unknown until observed or measured. The measurement of a physical state in quantum mechanics is different as it includes intrinsic uncertainty, which cannot be influenced by improvements in experimental techniques. This concept, originally proposed by Heisenberg, is called the uncertainty principle, which states that the uncertainties of measurement of energy ΔE and interval of time Δt, during which a microscopic particle possesses that energy:

$$\Delta E.\Delta t \geq h \qquad (7.4)$$

where h is Planck's constant. Hence, only the probabilities of getting particular results can be obtained in quantum mechanical experiments, fundamentally distinguishing them from the classical ones. The uncertainties in quantum-mechanical measurements stem from the disturbance which measurement itself causes to the measured state at a microscopic scale.

7.3.3   Quantum Superposition

There is an equal probability that something is either in one state (1) or another (0). Thus, something is in both states, or between both states at the same time until observed. Quantum superposition in qubits can be explained by flipping a coin. We know that the coin will land in one of two states: heads or tails. This is how binary computers think. While the coin is still spinning in the air, assuming your eye is not quick enough to "observe" the actual state it is in, the coin is actually in both states at the same time. Essentially until the coin lands it has to be considered both heads and tails simultaneously. A quantum computer manipulates qubits by executing a series of quantum gates, each being unitary transformation acting on a single qubit or pair of qubits [16, 17]

7.3.4   Quantum Entanglement (Nonlocality)

When two particles share the same quantum state they are entangled. This means that two or more particles will share the same properties: for example, their spins are related. Even when removed from each other, these particles will continue to share the same properties. By cleverly using quantum superposition and quantum entanglement, the quantum computer can be far more efficient than the classical computer [18–20]

7.4   Quantum Challenges

Challenges arise both from the difficulty in control and manipulation of quantum states, something that makes encoding quantum information difficult, as well as the vulnerability of that information to disturbance from the environment.

7.5   DNA Computing: From Bits to Cells

7.5.1   What Is DNA?

All organisms on this planet are made of the same type of genetic blueprint. Within the cells of any organism is a substance called deoxyribo nucleic acid (DNA) which is a double-stranded helix of nucleotides. DNA carries the genetic information of a cell. This information is the code used within cells to form proteins and is the building block upon which life is formed. Strands of DNA are long polymers of millions of linked nucleotides. The two strands of a DNA molecule are antiparallel where each strand runs in an opposite direction. DNA computing is a parallel computing model based on DNA molecule. The core reaction of DNA computing is the hybridization reaction between DNA molecules. The accuracy of the reaction between molecules directly affects the results of DNA computing [21–23].

7.5.2   Why DNA Computing?

DNA Computing (aka molecular computing or biological computing) is utilizing the property of DNA for massively parallel computation. With an appropriate setup and enough DNA, one can potentially solve huge problems by parallel search. Utilizing DNA for this type of computation can be much faster than utilizing a conventional computer.

7.5.3 How DNA Works?

DNA computing means the use of biological molecules, primarily deoxyribonucleic acid (DNA), DNA analogs, and (ribonucleic acid) RNA, for computation purpose. It is a completely new method of general computation alternative to semiconductor technology which uses biochemical processes based on DNA. Compared to the fastest supercomputer, DNA computer can achieve 10^6 operation/sec vs. 10^{14} operation/sec for DNA computer. Table 7.3 shows a comparison between DNA-based computer and conventional computer. Cells use chemicals for information storage and transfer while computers use magnetic or electronic means. In cells, proteins act as both programs and machines. In computers, programs and machines are separate with programs generally running the machines. Proteins contain instructions for self-assembly, computers don't.

Basic building blocks on DNA computing are shown in Fig. 7.5, where in every cell's nucleus, genes consist of tight coils of DNA's double helix. Number of genes/length of DNA depends on species. DNA may be viewed as logic memory or gate

Table 7.3 Comparison between DNA-based computer and conventional computer

DNA-based computers	Conventional (Si-based) computers
Slow at individual operations	Fast at individual operations
Can do billions of operations simultaneously	Can do substantially fewer operations simultaneously
Can provide huge memory in small space	Smaller memory
Setting up a problem may involve considerable preparations	Setting up only requires keyboard input
DNA is sensitive to chemical deterioration	Electronic data are vulnerable but can be backed up easily

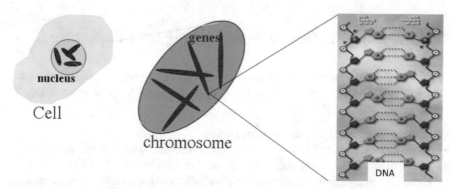

Fig. 7.5 Basic building blocks of DNA computing. The DNA is a double-stranded molecule. Each **strand** is based on four bases (Nucleic Acids): Adenine (**A**), Thymine (**T**), Cytosine (**C**), and Guanine (**G**). Those bases are linked through a sugar (deoxyribose). The nucleotide "A" pairs with "T" and "C" pairs with "G." Chromosomes are long strings of genes. Genes act as the functioning units of DNA molecules

AND actually it is a dense storage memory. DNA is a physical **chemical** system. Instructions are *coded* in a sequence of the DNA bases. A segment of DNA is exposed, transcribed, and translated to carry out instructions. DNA itself does not carry out any computation. It rather acts as a massive memory. BUT, the way complementary bases react with each other can be used to compute things [24, 25].

The DNA is a double-stranded molecule. Each strand is based on four bases (Nucleic Acids): Adenine (A), Thymine (T), Cytosine (C), and Guanine (G). Those bases are linked through a sugar (deoxyribose).

Sequencers are molecule-to-text converters. The extracted text sequence identifies the so-called primary structure of the measured molecule. DNA is being used because of its vast parallelism, energy efficiency, and the amount of information that the DNA can store. The available DNA sequences are about **163 million**. Sequence can be selected from these available DNA sequences. DNA plays a vital role in human structure as it determines the characteristic of a human. Any information can be represented in the form of DNA.

DNA Cryptography is a branch of DNA Computing. The cryptography approaches can be applied through DNA bases. It is a technique which converts the plaintext message into DNA strand. DNA storage can also be used as a storage element.

7.5.4   Disadvantages of DNA Computing

- High cost is time.
- Simple problems are solved much faster on electronic computers.
- It can take longer to sort out the answer to a problem than it took to solve the problem.
- Reliability as there is some time errors in the pairing of DNA strands.

7.5.5   Traveling Salesman Problem Using DNA-Computing

A salesman must find his way from city 0 to city 6, passing through each of the remaining cities only once, 7 nodes and 14 edges. This problem can be solved by the following algorithm [26]:

- Generate Random paths through graph G.
- From all paths created in step 1, keep only those that start at 0 and end at 6.
- From all remaining paths, keep only those that visit exactly 7 vertices.
- From all remaining paths, keep only those that visit each vertex at least once.
- If any path remains, return "yes"; otherwise, return "no".

Using DNA the algorithm can be as follows:

- Represent Each City By A DNA Strand of 20 Bases. For example:

Fig. 7.6 Traveling
salesman problem. Correct
path is marked in red

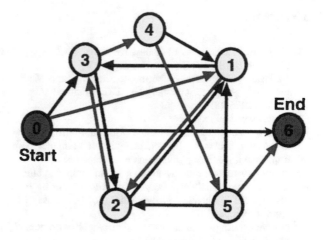

City 1 = ATGCTCAGCTACTATAGCGA

City 2=TGCGATGTACTAGCATATAT

Represent Each Air Route By Mixed Complementary Strands. For example:

City 1->2   **TGATATCGCTACGCTACATG**

- Mix the City DNA with the Path DNA. For example:

  {city1, city2, City 1->2} =
ATGCTCAGCTACTATAGCGATGCGATGTACTAGCATATA**TTGATATCGCTACGCTACATG**

After mixing they will form (7–2)! Different long DNA molecules. Select DNA
molecules with the right start and ends to determine the best pathway (defined by
the DNA sequence) (Fig. 7.6).

7.6   Conclusions

We are at the edge of a new age of computing. New results in quantum computing
have brought it closer to realization. Quantum computing is a promising trend to
overcome Moore's law limitations. DNA computing is still in its early stage, but it
is expected to be the future of data storage.

References

1. K. Salah, Y. Ismail, A. El-Rouby, *Arbitrary Modeling of TSVs for 3D Integrated Circuits* (Springer, Berlin, 2014)
2. National Academies of Sciences, Engineering, and Medicine, Quantum computing: progress and prospects (Washington, DC: National Academies Press, 2018). https://doi.org/10.17226/25196
3. http://www.qubit.org
4. http://www.cs.caltech.edu/~westside/quantum-intro.html
5. http://computer.howstuffworks.com/quantum-computer1.htm
6. http://en.wikipedia.org/wiki/Quantum_computers
7. http://www.carolla.com/quantum/QuantumComputers.htm
8. L. Vandersypen, *A "Spins-inside" Quantum Processor. Invited talk at PQCrypto 2017*, June 26–28, 2017, Utrecht, Netherlands (2017). https://2017.pqcrypto.org/conference/slides/vandersypen.pdf
9. http://www.cm.ph.bham.ac.uk/scondintro/qubitsintro.html
10. http://en.wikipedia.org/wiki/Qubits
11. http://www.csr.umd.edu/csrpage/research/quantum/index.html
12. M. Moller, C. Vuik, On the impact of quantum computing technology on future developments in high-performance scientific computing. Ethics Inf. Technol. **19**(4), 253–269 (2017). https://doi.org/10.1007/s10676-017-9438-0
13. T. Xin, S. Wei, J. Cui, J. Xiao, I. Arrazola, L. Lamata, X. Kong, D. Lu, E. Solano, G. Long, A quantum algorithm for solving linear differential equations: Theory and experiment. arXiv preprint arXiv **1807**, 04553 (2018)
14. K. Svore, M. Troyer, The quantum future of computation. Computer **49**(9), 21–30 (2016)
15. P. Ball, *Quantum Computer Simulates Hydrogen Molecule*. Chemistry World (2016). Accessed 31 Oct 2019
16. C. Day, Quantum computing is exciting and important–really! Comp Sci Eng **9**(2), 104–104 (2007)
17. L. Gyongyosi, S. Imre, A survey on quantum computing technology. Comp Sci Rev **31**, 51–71 (2019)
18. M.A. Nielsen, I.L. Chuang, *Quantum Computation and Quantum Information*, 2nd edn. (Cambridge University Press, Cambridge, 2010)
19. N.D. Mermin, *Quantum Computer Science* (CUP, Cambridge, 2007)
20. A.Y. Kitaev, A.H. Shen, M.N. Vyalyi, *Classical and Quantum Computation* (AMS, USA, 2002)
21. Z.F. Qiu, M. Lu, *Take Advantage of the Computing Power of DNAcomputers, in International Parallel and Distributed Processing Symposium* (Springer, Berlin, 2000), pp. 570–577
22. P.W.K. Rothemund, Folding DNA to create nanoscale shapes andpatterns. Nature **440**(7082), 297–302 (2006)
23. D. Limbachiya, M.K. Gupta, V. Aggarwal, Family of constrainedcodes for archival DNA data storage. IEEE Commun. Lett. **22**(10), 1972–1975 (Oct. 2018)
24. J. Bornholt, R. Lopez, D.M. Carmean, L. Ceze, G. Seelig, K. Strauss, *A DNA-Based Archival Storage System*, in Proc. 21st Int. Conf. Archit. Support Program. Lang. Oper. Syst. (ASPLOS), (2016), pp. 637–649
25. S.M.H.T. Yazdi, Y. Yuan, J. Ma, H. Zhao, O. Milenkovic, A rewritable, random-access DNA-based storage system. Sci. Rep. **5**(1), 1763 (2015)
26. J.Y. Lee, S.-Y. Shin, T.H. Park, B.-T. Zhang, Solving traveling salesman problems with DNA molecules encoding numerical values. Biosystems **78**, 39–47 (2004). https://doi.org/10.1016/j.biosystems.2004.06.005
27. K.S. Mohamed, Work around Moore's law: Current and next generation technologies, in *Applied Mechanics and Materials*, vol. 110, (Trans Tech Publications, Zurich, 2012), pp. 3278–3283

Chapter 8
Cloud, Fog, and Edge Computing

8.1 Cloud Computing

Cloud computing virtualizes the computing, storage, and band width. Thus, it reduces the deployment cost for software services and provides support for industrialization.

Cloud computing means storing and accessing data and programs over the Internet instead of your computer's hard drive. The cloud is just a metaphor for the Internet. Cloud computing is a shared pool of computing/storage resources that can be accessed on demand and dynamically offered to the user. Cloud computing services can be accessed at any time from any place. They are offered by many companies such as Google (AWS) and Microsoft (Azure). Moreover, there are open platforms for cloud computing such as Thingspeak and Thingsboard. Today, Amazon Web Services (AWS) is one of the leading cloud computing platforms in the industry [1].

Cloud service offerings are divided into three categories: infrastructure as a service (**IaaS**), platform as a service (**PaaS**), and software as a service (**SaaS**). IaaS is responsible for managing the hardware, network, and other services. PaaS supports the OS and application platform, and SaaS supports everything. Cloud computing services are shown in Fig. 8.1.

- **Infrastructure as a service (IaaS)**: You specify the low-level details of the virtual server you require, including the number of CPUs, RAM, hard disk space, networking capabilities, and operating system. The cloud provider offers a virtual machine to match these requirements. In addition to virtual servers, the definition of IaaS includes networking peripherals such as firewalls, load balancers, and storage. Therefore provisioning multiple load-balanced Java application servers on the cloud as well as storing your files on a cloud-based disk would both come under the IaaS model.
- **Platform as a service (PaaS)**: You choose between combinations of infrastructure and preconfigured software that best suits your needs. The cloud provider

© Springer Nature Switzerland AG 2020
K. S. Mohamed, *Neuromorphic Computing and Beyond*,
https://doi.org/10.1007/978-3-030-37224-8_8

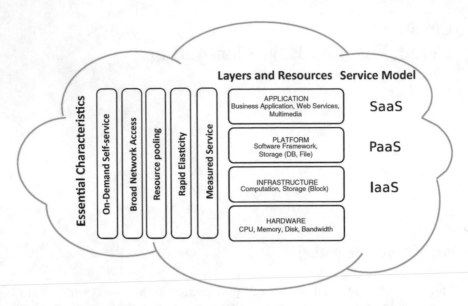

Fig. 8.1 Cloud computing services: an overall view

offers a virtual machine with preconfigured internal applications to match your requirements. The PaaS model is generally easier to use as you do not have to set up the underlying hardware and software; however, this can also be restrictive as your level of control on the underlying systems is significantly reduced compared to the IaaS model. The choice of infrastructure and pre-installed software differs between cloud providers, and if a cloud vendor does not provide something off the shelf that meets your needs, you are out of luck.

- **Software as a service (SaaS)**: You specify the kind of software application you want to use, such as a word processor. The cloud provider provisions the required infrastructure, operating system, and applications to match your requirement. Most SaaS cloud providers include a limited choice of the hardware characteristics that run the application and as a user you usually have no direct access to the underlying hardware that runs your application. You are also tied to the limitations of the hardware and software chosen by your cloud provider. If, for instance, a cinema chain is using a SaaS cloud-based booking system to manage ticket sales and the system is unavailable due to an internal error with the hardware used by the cloud provider, there is little the cinema chain can do but wait until the cloud provider has fixed the issue.

A cloud system may be public or private. Public clouds can be accessed by anyone. Private clouds offer services to a set of authorized users. A hybrid cloud is a combination of both public and private clouds [2–4].

- **Public cloud**: A public cloud provides services over the Internet to a consumer located anywhere in the world. The physical resources utilized by the provider to supply these services can also be anywhere in the world. This type of service could represent potential challenges to organizations such as banks that are prevented by regulatory requirements from storing confidential data on external systems. The cloud provider is generally responsible for procurement, setup, physical security, and maintenance of the physical infrastructure. To an extent the cloud provider is also responsible for the security of the application and data; this would depend on the service model (IaaS/PaaS/SaaS).

- **Private cloud**: A private cloud offers services to a single organization. Services are provided over a secure internal network and are not accessible to the general public over the Internet. The organization owns the physical hardware that supplies underlying services and is responsible for sctup, maintenance, and security of both the infrastructure and all the software that runs on the infrastructure. Because of the large infrastructure costs associated with this model, only very large corporations can afford to have their own private clouds. A private cloud is commonly referred to as an on-premises cloud, and can offer any combination of IaaS/SaaS/PaaS to the organization.
- **Hybrid cloud**: A hybrid cloud is essentially a cloud service that is composed of other types of cloud services. For example, a hybrid cloud could consist of both public and private clouds. The public sub-cloud could provide services that are intended for consumption by any user over the Internet. A private cloud could offer services that are sensitive to the business. Most large organizations use a hybrid cloud model. It lets them provision public cloud resources when needed for a specific business case, while continuing to enjoy the security and control that a private cloud allows for all their existing processes.

A cloud server is an open logical server which builds, hosts, and delivers through a cloud computing platform through the Internet. Cloud servers maintain and exhibit similar functionality as well as capabilities to a typical server. However, they are accessing remotely from a cloud service provider as open server. A cloud server known as a virtual private server or virtual server. A cloud server is an Infrastructure as a Service (IaaS) primarily based cloud service model [5]. Using cloud, computer resources are located anywhere with access over Internet and processing Data and information in a distributed mode. They provide sharing, archiving, and high-scale.

Cloud computing is a type of Internet-based computing that provides shared computer processing resources and data to computers and other devices on demand.

Cloud computing is a model for enabling ubiquitous, on-demand access to a shared pool of configurable computing resources (e.g., computer networks, servers, storage, applications, and services), which can be rapidly provisioned and released with minimal management effort.

The availability of virtually unlimited storage and processing capabilities at low cost enabled the realization of a new computing model, in which virtualized resources can be leased in an on-demand fashion, being provided as general utilities. Large companies (like Amazon, Google, and Facebook) widely adopted this paradigm for delivering services over the Internet, gaining both economic and technical benefits. Figure 8.2 shows a comparison between different types of cloud computing in terms of who manages?. Comparison between different cloud services is shown in Table 8.1.

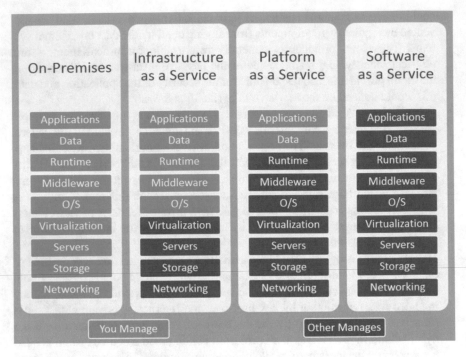

Fig. 8.2 Cloud computing services: who manages?

Cloud computing offers:

- Lower cost as Cloud computing eliminates the expense of setting up and running on-site data centers.
- Decrease the workforce.
- Focus mainly on app development instead of managing servers and focusing on maintenance.
- Higher security as Cloud service providers typically also offer a broad set of policies, compliance, technologies, and controls that strengthen your security posture by protecting your data, apps, and infrastructure from threats.

8.2   Fog/Edge Computing

Edge computing is a method of optimizing cloud computing systems by performing data processing at the edge of the network, near the source of the data. Recently, there has been a move towards system architecture, namely, fog computing. What this basically does is process the time-sensitive data on the edge of the network itself, i.e., closest to where the data is being generated so that appropriate actions can be taken in time. Some key features of this model are: minimize latency, conserve network bandwidth, resolve the data privacy and security issue, and increase reliability.

Table 8.1 Comparison between different cloud services

Point of comparison	AWS	Microsoft Azure	Google cloud
Caching	Elastic cache	Redis cache	Cloud CDN
Processor	In AWS, 128 can be the maximum processor in VM	In Azure, it can be 128	In Google cloud, it is only 96.
Marketplace	In this, AWS marketplace	Azure marketplace	G suite marketplace
App testing	In AWS, device farm is being used.	In Azure, DevTest labs are being used	Cloud test lab is being used in this.
GIT repositories	AWS source repositories	Azure source repositories.	Cloud source repositories.
Platform as service	Elastic beanstalk	Cloud services	Google app engine
Storage of object	S3	Block blob	Cloud storage
Managed data warehouse	Redshift	SQL warehouse	Big query
Kubernetes management	EKS	Kubernetes service	Kubernetes engine
File storage	EFS	Azure files	ZFS and Avere
Serverless computing	Lambda is being used for serverless computing	In Azure, Azure functions are used.	In Google cloud, cloud functions are used.
API management	Amazon API gateway	Azure API gateway	Cloud endpoints
Media services	Amazon elastic transcoder	Azure media services	Cloud video intelligence API
Website	Aws.amazon.com	Azure.microsoft.com	Cloud.google.com

Cloud computing is the computing paradigm that enables ubiquitous, convenient, on-demand network access to a shared pool of configurable computing resources (e.g., computing and storage facilities, applications, and services). Through virtualization technology, cloud computing shields the diversity of underlying devices and provides users with a variety of services in a transparent way, including IaaS (Infrastructure-as-a-Service), PaaS (Platform-as-a-Service), and SaaS (Software-as-a-Service). Due to the increasing number of access devices, cloud computing may face some problems in the bandwidth, latency, network unavailability, and security and privacy. Fog computing is considered as an extension of cloud computing to the edge network, providing services (e.g., compute, storage, networking) closer to near-user devices (e.g., network routers, various information systems), instead of sending data to cloud [6, 7].

Edge computing provides the opportunity to reduce latency when performing analytics in the cloud. But after data is collected on-premises, it is shared in the public cloud. Edge computing is going to be about having some on-premises compute that works with the public cloud in a hybrid way. Microsoft Azure Edge is an example of general-purpose edge computing. Clouds are composed of servers, where each server supports applications with computation and storage services. The device layer is composed of sensors and actuators. Sensor data collected by sensors is delivered to servers in networks. Sensor data is finally delivered to edge fog nodes

Fig. 8.3 Cloud computing paradigm

at the fog layer. Based on the sensor data, actions to be done by actuators are decided. Actuators receive actions from edge fog nodes and perform the actions on the physical environment. Fog nodes are at a layer between the device and cloud layers. Fog nodes are interconnected with other fog nodes in networks. A fog node supports the routing function where messages are routed to destination nodes, i.e., routing between servers and edge nodes like network routers. More importantly, a fog node does some computation on a collection of input data sent by sensors and other fog nodes. In addition, a fog node makes a decision on what actions actuators have to do based on sensor data. Then, the edge nodes issue the actions to actuator nodes. A fog node is also equipped with storages to buffer data. Thus, data and processes are distributed to not only servers but also fog nodes in the fog computing model while centralized to servers of clouds in the cloud computing model.

Data from various sources and domains is often data stream, such as numeric data from different sensors or social media text inputs. Common data streams generally follow the Gaussian distribution over a long period. However, data are produced in short time and in large quantities, presenting a variety of sporadic distributions over time. In addition, in some cases, in real time or near real time. One trend in Internet applications of Things that addresses the concept of Analytics is the use of Fog Computing that can decentralize the processing of data streams and only perform the transfer of filtered data from the devices from the edge of the network to the cloud. Therefore, the association of data stream analysis with Fog Computing allows companies to explore in real time the data produced and thus produce business value [8, 9]. Cloud computing paradigm versus edge computing paradigm is shown in Figs. 8.3 and 8.4, respectively [10, 11]. Comparison between cloud and fog are shown in Table 8.2 [12, 13].

Filling the technology gaps in supporting IoT will require a new architecture—Fog—that distributes computing, control, storage, and networking functions closer to end user devices along the cloud-to-things continuum.

Complementing the centralized cloud, Fog stands out along the following three dimensions:

- Carry out a substantial amount of data storage at or near the end user (rather than storing data only in remote data centers).
- Carry out a substantial amount of computing and control functions at or near the end user (rather than performing all these functions in remote data centers and cellular core networks).

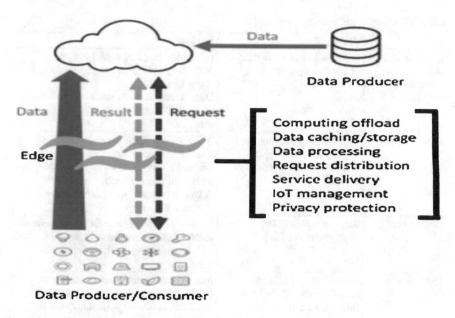

Fig. 8.4 Edge computing paradigm

- Carry out a substantial amount of communication and networking at or near the end user (rather than routing all network traffic through the backbone networks).

Fog interfaces are shown in Fig. 8.5, where Fog can interface to fog or cloud or user directly. Figure 8.6 shows another comparison between cloud/fog/edge computing.

8.3   Conclusions

This chapter gives insights into Cloud/Edge/Fog computing. Cloud computing is an example of an information technology (IT) paradigm, a model for enabling ubiquitous access to shared pools of configurable resources (such as computer servers, storage, applications and services, and networks), which can be rapidly provisioned with minimal management efforts, over the internet. Edge computing provides the opportunity to reduce latency when performing analytics in the cloud. But after data is collected on-premises, it is shared in the public cloud. Edge computing is going to be about having some on-premises compute that works with the public cloud in a hybrid way. Edge computing is a cutting-edge technology that, combined with cloud computing, will open up new dimensions in terms of flexibility, scalability, and data security and processing.

Table 8.2 Comparison between cloud and fog

Aspect	Cloud [15]	Fog [16, 17]
Location and model of computing	Centralized in a small number of big data centers	Often distributed in many locations, potentially over large geographical areas, closer to users. Distributed fog nodes and systems can be controlled in centralized or distributed manners.
Size	Cloud data centers are very large in size, each typically contain tens of thousands of servers.	A fog in each location can be small (e.g., one single fog node in a manufacturing plant or onboard a vehicle) or as large as required to meet customer demands. A large number of small fog nodes may be used to form a large fog system.
Deployment	Require sophisticated deployment planning.	While some fog deployments will require careful deployment planning, fog will enable ad hoc deployment with no or minimal planning.
Operation	Operate in facilities and environments selected and fully controlled by cloud operators. Operated and maintained by technical expert teams. Operated by large companies.	May operate in environments that are primarily determined by customers or their requirements. A fog system may not be controlled or managed by anyone and may not by operated by technical experts. Fog operation may require no or little human intervention. May be operated by large and small companies, depending on size.
Applications	Support predominantly, if not only, cyber-domain applications. Typically support applications that can tolerate round-trip delays in the order of a few seconds or longer.	Can support both cyber-domain and cyber-physical systems and applications. Can support significantly more time-critical applications that require latencies below tens of milliseconds or even lower.
Internet connectivity and bandwidth requirements	Require clients to have network connectivity to the cloud for the entire duration of services. Long-haul network bandwidth requirements grow with the total amount of data generated by all clients	Can operate autonomously to provide uninterrupted services even no or intermittent internet connectivity. Long-haul network bandwidth requirements grow with the total amount of data that need to be sent to the cloud after being filter by the fog

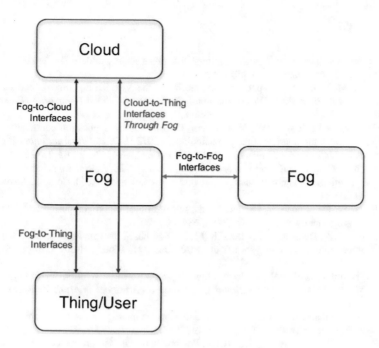

Fig. 8.5 Fog interfaces [14]

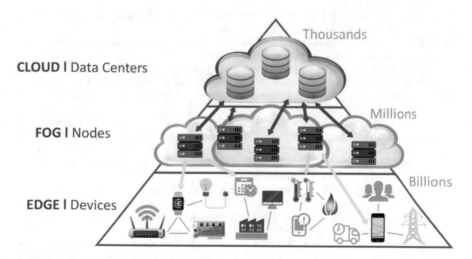

Fig. 8.6 Cloud vs. Fog vs. edge

References

1. K.S. Mohamed, The era of internet of things: Towards a smart world, in *The Era of Internet of Things*, (Springer, Cham, 2019), pp. 1–19
2. M. Mao, M. Humphrey, A performance study on the VM Startup time in the cloud, in *Proceedings of 2012 IEEE 5th International Conference on Cloud Computing (Cloud2012)* (2012), p. 423
3. S. He, L. Guo, Y. Guo, C. Wu, M. Ghanem, R. Han, Elastic application container: A light-weight approach for cloud resource provisioning, in *2012 IEEE 26th International Conference on Advanced Information Networking and Applications (AINA)* (2012), pp. 15–22
4. M. Netto, R. Calheiros, E. Rodrigues, R. Cunha, R. Buyya, HPC cloud for scientific and business applications: Taxonomy, vision, and research challenges. ACM Comput. Surv. **51**(1), 8:1–8:29 (2018)
5. A. Botta, W. De Donato, V. Persico, et al., Integration of cloud computing and internet of things. Futur. Gener. Comput. Syst. **56**(C), 684–700 (2016)
6. H.R. Arkian, A. Diyanat, A. Pourkhalili, MIST: Fog-based data analytics scheme with cost-efficient resource provisioning for IoT crowdsensing applications. J. Netw. Comput. Appl. **82**, 152–165 (2017)
7. J.B.M. Numhauser, XMPP Distributed topology as a potential solution for fog computing, in *MESH 2013 the Sixth International Conference on Advances in Mesh Networks* (2013), pp. 26–32
8. J. Pan, J. McElhannon, Future edge cloud and edge computing for internet of things applications. IEEE Internet Things J. **5**(1), 439–449 (2018)
9. K. Bilal, O. Khalid, A. Erbad, S.U. Khan, Potentials, trends, and prospects in edge technologies: Fog, cloudlet, Mobile edge, and micro data centers. Comput. Netw. **130**, 94–120 (2018)
10. W. Shi, J. Cao, Q. Zhang, et al., Edge computing: Vision and challenges. IEEE Internet Things J. **3**(5), 637–646 (2016)
11. T. Taleb, A. Ksentini, P. Frangoudis, Follow-me cloud: When cloud services follow Mobile users. IEEE Trans. Cloud Comp. **PP**(99), 1 (2016)
12. T. Taleb, S. Dutta, A. Ksentini, M. Iqbal, H. Flinck, Mobile edge computing potential in making cities smarter. IEEE Commun. Mag. **55**(3), 38–43 (2017)
13. T. Chakraborty, S.K. Datta, Home automation using edge computing and Internet of Things, in *2017 IEEE International Symposium on Consumer Electronics (ISCE)* (2017), pp. 47–49
14. M. Chiang, T. Zhang, Fog and IoT: An overview of research opportunities. IEEE Internet Things J. **3**, 854–864 (2016)
15. Byrne, J.; Svorobej, S.; Giannoutakis, K.M.; Tzovaras, D.; Byrne, P.J.; Östberg, P.; Gourinovitch, A.; Lynn, T., A review of cloud computing simulation platforms and related environments, in *Proceedings of the 7th International Conference on Cloud Computing and Services Science, Porto, Portugal, 24–26 April 2017* (SciTePress, Setubal, 2017), 1; pp. 679–691. https://doi.org/10.5220/0006373006790691
16. R. Buyya, R. Ranjan, R.N. Calheiros, Modeling and simulation of scalable Cloud computing environments and the CloudSim toolkit: Challenges and opportunities, in *Proceedings of the International Conference on IEEE High Performance Computing & Simulation (HPCS'09), Leipzig, Germany*, 21–24 June 2009, pp. 1–11
17. Y. Li, A.-C. Orgerie, I. Rodero, B. Lemma Amersho, M. Parashar, J.-M. Menaud, End-to-end energy models for edge cloud-based IoT platforms: Application to data stream analysis in IoT. Futur. Gener. Comput. Syst. **87**, 667–678 (2018)

Chapter 9
Reconfigurable and Heterogeneous Computing

9.1 Embedded Computing

Embedded system is anything that uses a microprocessor but is not a general-purpose computer such as televisions, video games, refrigerators, cars, planes, elevators, remote controls, alarm systems, printers, and scanners. The end user sees a smart system as opposed to the computer inside the system, but he does not or cannot modify or upgrade the internals. Embedded systems usually consist of hardware and software. Embedded systems must be efficient in terms of energy, code-size, run-time, weight, and cost. An embedded system is any device controlled by instructions stored on a chip. These devices are usually controlled by a microprocessor that executes the instructions stored on a read-only memory (ROM) chip. The software for the embedded system is called firmware. Embedded systems are also known as real-time systems since they respond to an input or event and produce the result within a guaranteed time period [1].

9.1.1 Categories of Embedded Systems Are [2–5]

- General Computing: Applications similar to desktop computing, but in an embedded package such as Video games, wearable computers, and automatic tellers.
- Control Systems: Closed loop feedback control of real-time system such as vehicle engines, chemical processes, nuclear power, and flight control.
- Signal Processing: Computations involving large data streams such as Radar, Sonar, and video compression.
- Communication & Networking: Switching and information transmission, telephone system, Internet.

© Springer Nature Switzerland AG 2020
K. S. Mohamed, *Neuromorphic Computing and Beyond*,
https://doi.org/10.1007/978-3-030-37224-8_9

9.1.2 Embedded System Classifications

Stand-alone Embedded systems:

- Works by itself. It is a self-contained device which does not require any host system like a computer, e.g., temperature measurement systems, MP3 players, digital cameras, and microwave ovens.

Real-time embedded systems:

- Gives the required output in a specified time (deadline).

Soft Real-Time system:

- Violation of time constraints will cause only the degraded quality, but the system can continue to operate, e.g., washing machine, TV remote.

Hard Real-Time system:

- Violation of time constraints will cause critical failure and loss of life or property damage or catastrophe, e.g., deadline in a missile control embedded system, delayed alarm during a gas leakage, car airbag control system.

Networked embedded systems:

- Related to a network with network interfaces to access the resources. The connected network can be a LAN or a WAN, or the Internet. The connection can be either wired or wireless, e.g., home security system.

Mobile Embedded systems

- Mobile and cellular phones, digital cameras, MP3 players, and PDA. Limitation is memory and other resources.

  Small-scaled embedded system: Supported by a single 8–16 bit Microcontroller with on-chip RAM and ROM.
  Medium-scaled embedded system: Supported by 16–32 bit Microcontroller/Microprocessor with external RAM and ROM.
  Large-scaled embedded system: Supported by 32–64 bit multiple chips which can perform distributed jobs.

9.1.3 Components of Embedded Systems

The layered architecture for embedded system consists of (Fig. 9.1):

- **Hardware:** Processor/controller, Timers, Interrupt controller, I/O Devices, Memories, Ports, etc.
- **Application Software:** Which may perform concurrently the series of tasks or multiple tasks.

Fig. 9.1 Layered
architecture for embedded
system

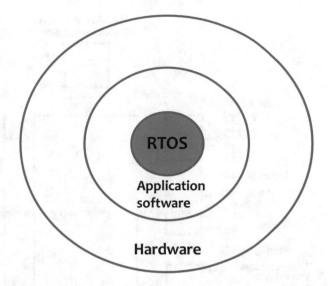

- **RTOS**: RTOS defines the way the system work. It supervises the application
  software. It sets the rules during the execution of the application program. A
  small scale embedded system may not need an RTOS.

9.1.4   Microprocessor vs. Microcontroller

The microprocessor is a processor on one silicon chip. The microcontrollers are
used in embedded computing. The microcontroller is a microprocessor with added
circuitry such as Timers, A/D converter, and Serial I/O as depicted in Fig. 9.2. There
are many families of microcontrollers (Table 9.1):

- 8051

  - Intel
  - Philips
  - Atmel
  - Siemens
  - Dallas

- Motorola
- PIC
- Hitachi
- Texas
- ARM
- Others

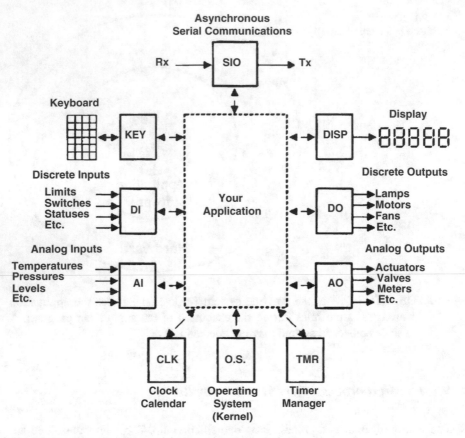

Fig. 9.2 Typical embedded system: Microcontroller based

9.1.5 Embedded Systems Programming

In terms of programming, each processor has its own C library and its own compiler. Sometimes, we can program it using assembly in case of high speed, and low code size is needed. However C is the most common language as it provides an additional level above **assembly** programming; it is fairly efficient and supports access to I/O. Moreover, it ensures ease of management of large embedded projects. Layers of computer system is shown in Fig. 9.3.

9.1.6 DSP

Digital signal processors (DSP) are processors targeted at and optimized for processing of digitally represented analog signals. Nearly all architectural features included in a DSP are associated with specific algorithms whose computation improves by the

Table 9.1 Comparison between different microcontrollers

	8051	PIC	AVR	ARM
Bus width	8-bit	8/16/32-bit	8/32-bit	32-bit mostly also available as 64-bit
Communication protocols	UART, USART, SPI, I2C	UART, USART, LIN, CAN, Ethernet, SPI, I2S	UART, USART, SPI, I2C	UART, USART, LIN, I2C, SPI, CAN, USB, Ethernet, I2S, DSP, SAI (serial audio interface), IrDA
Speed	12 Clock/ instruction cycle	4 Clock/ instruction cycle	1 Clock/ instruction cycle	1 Clock/instruction cycle
Memory	ROM, SRAM, FLASH	SRAM, FLASH	Flash, SRAM, EEPROM	Flash, SDRAM, EEPROM
Memory Architecture	Von-Neumann architecture	Harvard architecture	Modified Harvard architecture	Modified Harvard architecture
Power Consumption	Average	Low	Low	Low
Manufacturer	NXP, Atmel, Silicon Labs, Dallas, Cyprus, Infineon, etc.	Microchip	Atmel	Apple, Nvidia, Qualcomm, Samsung electronics, and TI etc.
Cost (as compared to features provided)	Very low	Average	Average	Low

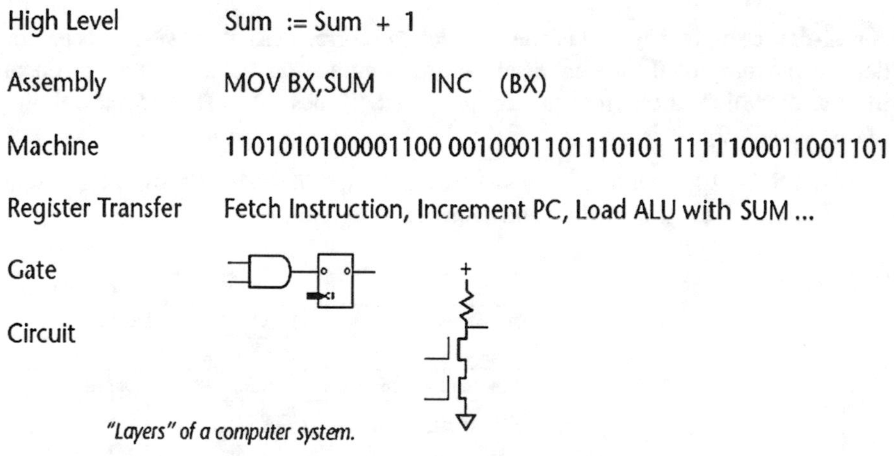

High Level Sum := Sum + 1

Assembly MOV BX,SUM INC (BX)

Machine 110101010000110 0 0010001101110101 1111100011001101

Register Transfer Fetch Instruction, Increment PC, Load ALU with SUM ...

Gate

Circuit

"Layers" of a computer system.

Fig. 9.3 Layers of computer system

addition of the feature. For applications that are still evolving, DSPs provide a middle-point in architecture design between a custom ASIC fixed function hardware block and general-purpose processors. Due to their specialization, DSPs provide high-performance programmable logic blocks to implement newer standards and yet are more efficient at execution compared to a CPU. DSPs are employed in a number of applications such as voice processing, audio processing, image processing, and video processing. DSPs support a fast single-cycle multiply-accumulate (MAC) instruction, since multiplication and a subsequent accumulation of products is a common operation in signal processing. A hardware feature called zero overhead loops provides an instruction to specify the loop count and the loop-start address, which enables low overhead loop execution without the need to update and test the loop counter or branching back to the loop-start instruction. Address generation units support several special addressing modes for efficient handling of predictable memory access patterns in DSP algorithms. Register-indirect addressing with post-increment automatically increments the address pointer and is useful for accessing data that is stored sequentially in memory. Circular addressing allows access to a sequential block of data followed by automatic wrap around to the starting address. Without these features, explicit instructions are required to update the address pointer. Finally, DSPs support variable data width and fixed precision data formats. Fixed point processing is cheaper and less power consuming than FP processing, and the variable data widths enable using the shortest data word that can provide the required accuracy for target applications, thus enabling cost-sensitive use-cases [6].

9.2   Real-Time Computing

A real-time computer system is one in which the correctness of the system behavior depends not only on the logical results of the computation, but also on the physical instant at which these results are produced. Types of RTS are as follows (Table 9.2) [7–9]:

- **Hard RTS**: tight limits on response time, so that a delayed result is a wrong result, e.g., jet fuel controller and camera shutter unit.

Table 9.2 Hard vs. Soft RTS

Mteric	Hard real time	Soft real time
Deadlines	Hard	Soft
Pacing	Environment	Computer
Peak load performance	Predictable	Degraded
Error detection	System	User
Safety	Critical	Noncritical
Redundancy	Active	Standby
Time granularity	Millisecond	Second
Data files	Small/medium	Large
Data integrity	Short-term	Long-term

- **Soft RTS**: need to meet only time-average performance target. As long as most results are available before deadline the system will run successfully. For example, audio and video transmission, single frame skip is fine, but repeated loss is unacceptable.
- **Firm RTS**: somewhere between the two, e.g., Space station solar panel unit.

Scheduling can be preemptive and non-preemptive scheduling: Non-preemptive schedulers are based on the assumption that tasks are executed until they are done. As a result, the response time for external events may be quite long if some tasks have a large execution time. Preemptive schedulers have to be used if some tasks have long execution times or if the response time for external events is required to be short. Moreover, scheduling can be dynamic or static scheduling. In dynamic scheduling, processor allocation decisions are done at run-time, while in static scheduling, processor allocation decisions are done at design-time. The time at which a real-time system has to produce a specific result is called a deadline.

Deadlines are dictated by the environment. Real-time concepts can be used in many applications [10].

9.3 Reconfigurable Computing

Everything can be reconfigured over and over at run time (Run-Time Reconfiguration) to suite underlying applications. A comparison between Microprocessor system vs. reconfigurable system (FPGA based) is shown in Fig. 9.4. A computer that can *RE*-configure itself to perform computation spatially as needed. Today most of the System on Chips (SoCs) is being realized by using higher end FPGAs as reconfigurable computing area and soft-core processors as software computing unit [11–13].

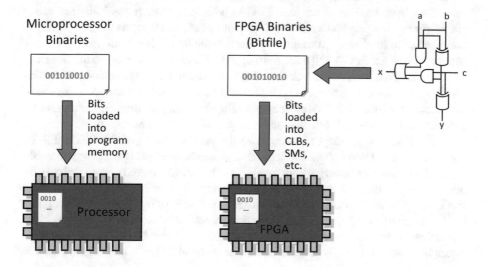

Fig. 9.4 Microprocessor system vs. reconfigurable system

9.3.1   FPGA

The advantage of FPGA-based systems over traditional processing units-based systems such as desktop computers, smartphones, and GPUs is the availability of freely programmable, general-purpose logic blocks. FPGAs can be arranged into high-performance specialized accelerators for very specific tasks, resulting in improved processing speed, higher throughput. Compared to GPUs, FPGAs are considered to be a much power-efficient device where they fit better for mobile device-based applications. These advantages come at the price of increased complexity and reduced agility during development time, where designers need to carefully take into consideration the available hardware resources and the efficient mapping of target algorithms onto the FPGA architecture. Further, FPGAs exceed the computing power of digital signal processors (DSPs) by breaking the paradigm of sequential execution and accomplishing more per clock cycle where they take full advantage of hardware parallelism. Controlling inputs and outputs (I/O) at the hardware level provides faster response time and specialized functionality to closely match application requirements. FPGAs usually do not use operating systems that actually minimize reliability concerns with true parallel execution and deterministic hardware that is dedicated to every task [14].

Application-Specific Integrated Circuits (ASICs) are custom-tailored semiconductor devices. Unlike FPGAs, ASICs do not have any area or timing overhead that could be caused by configuration logic and generic interconnects, thus resulting in the fastest, most energy-efficient, and smallest systems. However, the sophisticated fabrication processes for ASICs result in a very lengthy and complicated development round and very high nonrecurring engineering upfront costs that demand a first-time-right design methodology and very extensive design verification. Therefore, ASICs are mostly suited for very-high-volume, cost-sensitive applications where the nonrecurring engineering and fabrication costs can be shared between a large number of devices. FPGAs with their reprogram-ability are better suited for prototyping and short development cycles, where concepts can be tested and verified in hardware without going through the long fabrication process of custom ASIC design. FPGA chips are field upgradable and do not require the time and expense involved with ASIC redesign. Digital communication protocols, for example, have specifications that can change over time, and ASIC-based interfaces may cause maintenance and forward-compatibility challenges. Being reconfigurable, FPGAs can keep up with future modifications that might be necessary.

While it is best to adapt algorithms to the parallel nature of the GPUs, FPGA architecture is tailored for the application, they can be built using the programmable logic blocks to meet the algorithm needs. In other words, there is less emphasis on adapting algorithms when it comes to developing machine learning techniques for FPGAs. This allows more freedom to explore algorithm-level optimizations. The performance of FPGA design can be further increased which we will explore later. Optimization techniques that require many complex low-level hardware control operations cannot be easily implemented in high-level software languages; thus it is

Table 9.3 Performance comparison between FPGA and GPU

Feature	Winner
Floating-point processing	GPU
Timing latency	FPGA
Processing power	FPGA
Interfaces	FPGA
Development ease	GPU
Reuse	GPU
Size	FPGA

more attractive to consider FPGA implementation. Further, FPGAs are reconfigurable and flexible and thus offer a wide scope of CNN models to be implemented on the same chip without spending any further design costs as it is the case in ASICs. Thus far, FPGA is the most suitable platform for our algorithm. Performance comparison between FPGA and GPU is shown in Table 9.3.

9.3.2 High-Level Synthesis (C/C++ to RTL)

With HLS, designers can implement their designs through loops, arrays, floats, function calls, and other relevant arithmetic operations. The used loops, arrays, function calls, etc. are converted into counters, multiplexers, multipliers, memories, computation cores, and handshake protocols. The compilation can be guided using scripted compiler directives or compiler pragmas, which are meta-instructions interpreted directly by the HLS compiler. Vivado High-level Synthesis (Vivado HLS), offered by Xilinx, is the most common commercial HLS tool. Although HLS can provide faster development cycles, easier track for hardware implementation, and higher productivity, HLS tools do not provide sufficient optimization for a lot of applications. Optimization in HLS is limited and defined by the directives and programs that are embedded in the tool. As a matter of fact, HLS tools have been on the market for about years now, yet designers still use hardware description languages for their FPGA designs. The task of converting sequential, high-level software descriptions into fully optimized, parallel hardware architectures is extremely complex. Although companies have invested hundreds of millions of dollars and years of research into HLS, the results attained are still highly dependent on the coding style and intricate design details. Because flaws and deficiencies in the compiler are only discovered during the design, the decision for HLS is associated with a non-negligible risk. Having said that the implementation of algorithms using HDLs is tedious and complicated and optimization levels are not met using HLS, designers find themselves bound and have to trade off optimization for development round or vice versa (Fig. 9.5).

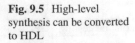

Fig. 9.5 High-level
synthesis can be converted
to HDL

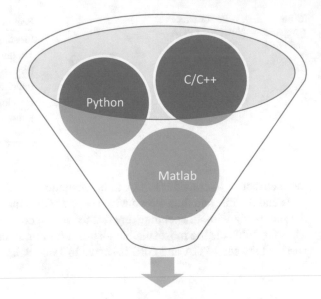

VHDL/Verilog

9.3.3 High-Level Synthesis (Python to HDL)

MyHDL turns Python into a hardware description and verification language, providing hardware engineers with the power of the Python ecosystem [15].

MyHDL is a free, open-source package for using Python as a hardware description and verification language. Python is a very-high-level language, and hardware designers can use its full power to model and simulate their designs. Moreover, MyHDL can convert a design to Verilog or VHDL. This provides a path into a traditional design flow.

Subject to some limitations, MyHDL supports the automatic conversion of MyHDL code to Verilog or VHDL code. This feature provides a path from MyHDL into a standard Verilog or VHDL-based design environment. To be convertible, the hardware description should satisfy certain restrictions, defined as the convertible subset.

A convertible design can be converted to an equivalent model in Verilog or VHDL, using the function to Verilog or to VHDL from the MyHDL library. The converter attempts to issue clear error messages when it encounters a construct that cannot be converted. Coding style a natural restriction on convertible code is that it should be written in MyHDL style: cooperating generators, communicating through signals, and with sensitivity specify resume conditions.

Supported types are the most important restriction regards object types. Only a limited amount of types can be converted. Python int and long objects are mapped to Verilog or VHDL integers. All other supported types need to have a defined bit width. The supported types are the Python bool type, the MyHDL int by type, and

MyHDL enumeration types returned by function Enum. The known issue is that Verilog and VHDL integers are 32-bit wide. Python is moving towards integers with undefined width. Python int and long variables are mapped to Verilog integers; so for values wider than 32 bit this mapping is incorrect. Synthesis pragmas are specified as Verilog comments. Inconsistent place of the sensitivity list is inferred from always comb. The semantics of always comb, both in Verilog and MyHDL, is to have an implicit sensitivity list at the end of the code. However, this may not be synthesizable. Therefore, the inferred sensitivity list is put at the top of the corresponding always block or process. This may cause inconsistent behavior at the start of the simulation. The workaround is to create events at time 0.

9.3.4 MATLAB to HDL

HDL Coder generates portable, synthesizable Verilog and VHDL code from MATLAB functions, Simulink® models, and State flow charts. The generated HDL code can be used for FPGA programming or ASIC prototyping and design.

IIDL Coder provides a workflow advisor that automates the programming of Xilinx, Micro semi, and Intel FPGAs. You can control HDL architecture and implementation, highlight critical paths, and generate hardware resource utilization estimates.

Vision HDL Toolbox provides pixel-streaming algorithms for the design and implementation of vision systems on FPGAs and ASICs. It provides a design framework that supports a diverse set of interface types, frame sizes, and frame rates, including high-definition (1080p) video. The image processing, video, and computer vision algorithms in the toolbox use an architecture appropriate for HDL implementations. The toolbox algorithms are designed to generate readable, synthesizable code in VHDL and Verilog (with HDL Coder™). The generated HDL code can process 1080p60 in real time. Intellectual property (IP) blocks in Vision HDL Toolbox provide efficient hardware implementations for computationally intensive streaming algorithms that are often implemented in hardware, enabling you to accelerate the design of image and video processing subsystems.

9.3.5 Java to VHDL

Because it is impractical to implement CNN models, especially large ones using HDL from scratch, we present a VHDL generation tool (VGT) based on Java language that offers a parameterized implementation to overcome the barriers introduced by high description languages and the limitations of HLS tools.

The generated implementation is fully pipelined, each stage in the design is properly pipelined to hide latency; highly parallel, parallelism is highly utilized corresponding

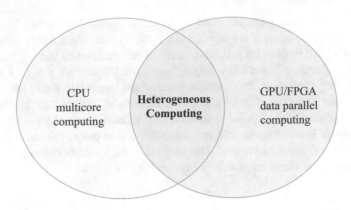

Fig. 9.6 Heterogeneous computing

to the available hardware resources; scalable and reconfigurable, the implementation can easily be reconfigured either using VGT or directly using the generated VHDL code in accordance to desired changes; and modular, the generated code is broken down into multiple VHDL modules.

The tool run validation checks on imported parameters to verify if they match up with the configured model and parameters representation. If the parameters file content violates any of the aforementioned rules, then the file will not be imported/loaded to the tool and an information message will be displayed to the user stating what errors should be fixed. Upon successful parameters inclusion, code generation module is enabled and users can generate VHDL code for their model as well as a test-bench for simulation and validation purposes.

9.4 Heterogeneous Computing

Heterogeneous system is a system made up of architecturally diverse user-programmable computing units. It is typically a mixture of two or more of the following: RISC/CISC CPUs, Graphical Processing Units (GPU), and Reconfigurable Computing Elements [16]. Heterogeneous computing increases computational power (Fig. 9.6). Moreover, we can integrate both CPU and GPU into a single chip by removing PCI-e bus to reduce the data transfer cost between CPU and GPU (Fig. 9.7). It is also called multiprocessor system-on-chips (MPSoC). There are many software challenges for heterogeneous computing. The exponential growth in data is already outpacing both the processor and storage technologies. Moreover, with the current semiconductor technology hitting its physical limitation, it is heterogeneous computing that is becoming the new norm to cater to the surging computational demands in a power-efficient manner. For example, various specialized processing architectures, or coprocessors, such as GPUs, TPUs, FPGAs, and ASICs not only have different tradeoffs (e.g., computing power, parallelism, memory bandwidth, and power consumption) but also are being proposed, developed, and

Fig. 9.7 Combining CPU
and GPU into a single chip

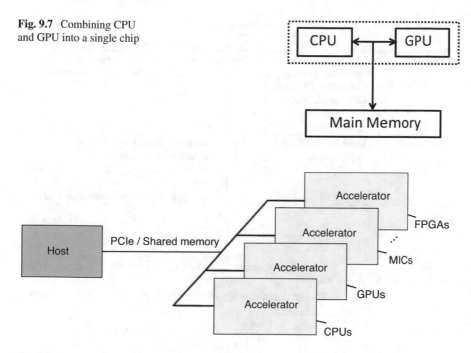

Fig. 9.8 Heterogeneous platforms

commercialized (Fig. 9.8) [17, 18]. Heterogeneous computing platforms can be found in every domain of computing—from high-end servers and high-performance computing machines all the way down to low-power embedded devices including mobile phones and tablets.

9.4.1 *Heterogeneity vs. Homogeneity*

- Increased performance as both devices work in parallel.
- Decreased data communication which is often the bottleneck of the system.
- Different devices for different roles.
- Increased flexibility and reliability.
- Increased power efficiency.

9.4.2 *Pollack's Rule*

Performance increase is roughly proportional to the square root of the increase in complexity as shown in Eq. (9.1) and power consumption increase is roughly linearly proportional to the increase in complexity as shown in Eq. (9.2).

Table 9.4 Comparison between static and dynamic partitioning

	Static partitioning	Dynamic partitioning
Advantages	• Can be computed before runtime ≥no overhead. • Can detect GPU-only/ CPU-only cases. • No unnecessary CPU-GPU data transfers.	• Responds to runtime performance variability works for all applications.
Disadvantages	• Does not work for all applications.	• Might introduce high CPU-GPU data-transfer overhead. • Might not work for CPU-only/GPU-only cases. • Incurs high runtime scheduling overhead.

$$\text{Performance} \propto \sqrt{\text{complexity}} \qquad\qquad (9.1)$$

$$\text{Power consumption} \propto \text{complexity} \qquad\qquad (9.2)$$

9.4.3 Static vs. Dynamic Partitioning

Comparison between static and dynamic partitioning is shown in Table 9.4.

9.4.4 Heterogeneous Computing Programming

With the rapid development of multicore technology, the number of cores in CPU has been increasing. The CPUs with 4-cores, 6-cores, 8-cores, and more cores enter the general computing environment to improve rapidly the parallel processing power. A heterogeneous computing environment can be built up with GPU and multicore CPU. The GPU does not have a process control capability as a device in CUDA, which is controlled by CPU. The data are transported from host memory to the global memory of GPU. Then CPU invokes the calculation process of GPU by calling the kernel function. OpenMP provides a simple and easy-to-use parallel computing capability of multi-threading on multicore CPUs. A heterogeneous programming model can be established by combining OpenMP and CUDA for a CPU–GPU heterogeneous computing environment. OpenMP dedicates one thread for controlling the GPU, while the other threads are used to share the workload among the remaining CPU cores initially, the data must be divided into two sets which are assigned to CPU and GPU, respectively. Then, two groups of threads are created in the parallel section of OpenMP, where a single thread is dedicated to controlling the GPU while other threads undertake the CPU workload by utilizing the remaining CPU cores [19, 20].

9.4.4.1 Heterogeneous Computing Programming: OpenCL

Open Computing Language is a framework for writing programs that execute across heterogeneous platforms consisting of central processing units (CPUs), graphics processing units (GPUs), digital signal processors (DSPs), field-programmable gate arrays (FPGAs) and other processors or hardware accelerators.

The traditional FPGA design methodology using hardware description language (HDL) is inefficient because it is becoming difficult for designers to develop details of circuits and control states for large and complex FPGAs. Design methodologies using C language are more efficient than HDL-based ones, but it is still difficult to design a total computing system. The C-based design tools can generate only the data path inside FPGAs and do not support interface between FPGAs and external memory devices or interface between FPGAs and host CPUs. Designers still need to use HDL to design interface circuits. To solve these problems, a design environment based on Open Computing Language (OpenCL) has recently been brought to FPGA design. OpenCL is a C-based design environment for a heterogeneous computing platform consisting of a host CPU and accelerators, such as GPUs and FPGAs. OpenCL allows designers to develop whole computation: computation on the host, data transfer between the host and accelerators, and computation on accelerators. Hence, by analyzing OpenCL codes, the OpenCL design environment for FPGAs can generate FPGA circuits together with interface circuits [21, 22].

9.5 Conclusions

In this chapter, embedded computing, reconfigurable computing, and real-time computing are introduced. Moreover, heterogeneous computing is discussed. Heterogeneous system is a system made up of architecturally diverse user-programmable computing units. It is typically a mixture of two or more of the following: RISC/CISC CPUs, Graphical Processing Units (GPU), and Reconfigurable Computing Elements.

References

1. M. Wolf, *Computers as Components: Principles of Embedded Computing System Design*, 4th edn. (Morgan Kaufmann, Burlington, 2017)
2. M. Barr, *Embedded Systems Glossary* (Neutrino Technical Library)
3. S. Heath, *Embedded Systems Design.* EDN series for design engineers (2nd ed.) (2003) Amsterdam: Elsevier
4. C. Alippi, *Intelligence for embedded systems* (Springer, Berlin, 2014), p. 283
5. S. Mittal, A survey of techniques for improving energy efficiency in embedded computing systems. IJCAET **6**(4), 440–459 (2014)
6. J. Eyre, J. Bier, The evolution of DSP processors. IEEE Signal Proc Mag **17**(2), 43–51 (2000). https://doi.org/10.1109/79.826411

7. A. Burns, A. Wellings, *Real-Time Systems and Programming Languages*, 4th edn. (Addison-Wesley, Boston, 2009)
8. G. Buttazzo, *Hard Real-Time Computing Systems: Predictable Scheduling Algorithms and Applications* (Springer, New York, 2011)
9. J.W.S. Liu, *Real-Time Systems* (Prentice Hall, Upper Saddle River, 2000)
10. K. Salah, *Real Time Embedded System IPs Protection Using Chaotic Maps*. In: Ubiquitous computing, electronics and Mobile communication conference (UEMCON), 2017 IEEE 8th annual. IEEE (2017)
11. S. Hauck, A. DeHon, *Reconfigurable Computing: The Theory and Practice of FPGA-Based Computing* (Morgan Kaufmann, Burlington, 2008)
12. J. Henkel, S. Parameswaran (eds.), *Designing Embedded Processors. A Low Power Perspective* (Springer, Berlin, 2007)
13. J. Teich et al., Reconfigurable computing systems. Spec Top Iss J **49**(3) (2007)
14. K. Salah, *An Area Efficient Multi-mode Memory Controller Based on Dynamic Partial Reconfiguration*. In 2017 8th IEEE Annual Information Technology, Electronics and Mobile Communication Conference (IEMCON). IEEE (2017), pp. 328–331
15. http://www.myhdl.org/
16. A. Klöckner et al., PyCUDA and PyOpenCL: A scripting-based approach to GPU run time code generation. Parallel Comput **38**(3), 157–174 (2012)
17. S. Mittal, J. Vetter, A survey of CPU-GPU heterogeneous computing techniques. ACM Comput Surv (2015)
18. A. Venkat, D.M. Tullsen, *Harnessing ISA Diversity: Design of a Heterogeneous-ISA Chip Multiprocessor*. In: Proceedings of the 41st Annual International Symposium on Computer Architecture (2014)
19. W. Yang, K. Li, K. Li a, A hybrid computing method of SpMV on CPU–GPU heterogeneous computing systems. J Paral Distr Comp **104**, 49–60 (2017)
20. M. Kreutzer, G. Hager, G. Wellein, et al., *Sparse Eatrix–Vector Multiplication on GPGPU Clusters: A New Storage Format and a Scalable Implementation*, in: Proceedings of the 2012 IEEE 26th International Parallel and Distribute Processing Symposium Workshops & Ph.D. Forum, IPDPSW 12, (IEEE Comp Soc, Washington, DC, 2012), pp. 1696–1702
21. H. Muthumala, Waidyasooriya, M. Hariyama, K. Uchiyama, *Design of FPGA-Based Computing Systems with OpenCL* (Springer, Berlin, 2018)
22. J. Long, *Hands On OpenCL: An Open Source Two-Day Lecture Course for Teaching and Learning OpenCL* (2018), https://handsonopencl.github.io/. Accessed 25 Jun 2018

Chapter 10
Conclusions

In this book, the traditional computing concepts are introduced. The main walls and the need for new trends in computing are highlighted. New trends are needed to enhance computing capability as Moore's law is slowing down, but the demand for computing keeps increasing.

The numerical computing methods for electronics are introduced. Numerical analysis is the study of algorithms that use numerical approximation for the problems of mathematical analysis. Numerical analysis naturally finds application in all fields of engineering and the physical sciences. Also the life sciences, social sciences, medicine, business, and even the arts have adopted elements of scientific computations. Different approaches to solve partial differential equations (PDEs), ordinary differential equations (ODEs), system of nonlinear equations (SNLEs), system of linear equations (SLEs), as well as the advantages and disadvantages of each method are analyzed.

In this book, parallel computing is introduced. A performance comparison of MPI, Open MP, and CUDA parallel programming languages is presented. The performance is analyzed using cryptography as a case study.

Moreover, in this book a comprehensive guide introducing the concepts, the current status, the challenges and opportunities, as well as emerging applications of deep learning and cognitive computing is introduced. Deep learning (DL) can be applied to all types of data (test, audio, image, video). Deep learning and machine learning can improve quality of design automation in many ways.

Sacrificing exact calculations to improve performance in terms of run-time, power, and area is at the foundation of approximate computing. The survey shows that the approximate computing is a promising paradigm towards implementing ultra-low-power systems with an acceptable quality for applications that do not require exact results. Approximate computing exposes tradeoffs between accuracy and resource usage. It is important paradigm especially for resource-constrained embedded systems. Approximate computing is a promising technique to reduce energy consumption and silicon area. The approximate computing paradigm seeks to improve the efficiency of a computer system by reducing the quality of the

© Springer Nature Switzerland AG 2020
K. S. Mohamed, *Neuromorphic Computing and Beyond*,
https://doi.org/10.1007/978-3-030-37224-8_10

results. AC aims at relaxing the bounds of exact computing to provide new opportunities for achieving gains in terms of energy, power, performance, and/or area efficiency at the cost of reduced output quality, typically within the tolerable range.

Processing in memory (PIM) promises to solve the bandwidth bottleneck issues in today's systems. PIM enables moving from big data era to fast data era. PIM-enabled architectures achieve impressive performance results by integrating processing units into the memory. NV memory chips could lead the way to fast, energy-efficient hardware accelerators. PIM reduces data movement by performing Processing in Memory. DRAM memory controllers have gained a lot of popularity in recent years. They are widely used in applications ranging from smart phones to high-performance computers. These applications require large amount of memory accessing. However, memory-wall is still a bottleneck. In this chapter, we introduce the most recent techniques used to enhance DRAM memory controllers in terms of power, capacity, latency, bandwidth, and area. The traditional information system adopts the architecture of separation of computing and storage. The data is transmitted between the CPU and the memory, and the power consumption is large and the speed is slow. The PIM adopts the architecture of storage and computing together, which eliminates the data transmission process and greatly improves the information processing efficiency.

We are at the edge of a new age of computing. New results in quantum computing have brought it closer to realization. Quantum computing is a promising trend to overcome Moore's law limitations. DNA computing is still in its early stage, but it is expected to be the future of data storage.

This book gives insights into Cloud/Edge/Fog computing. Cloud computing is an example of an information technology (IT) paradigm, a model for enabling ubiquitous access to shared pools of configurable resources (such as computer servers, storage, applications and services, and networks), which can be rapidly provisioned with minimal management efforts, over the internet. Edge computing provides the opportunity to reduce latency when performing analytics in the cloud. But after data is collected on-premises, it is shared in the public cloud. Edge computing is going to be about having some on-premises compute that works with the public cloud in a hybrid way.

Index

© Springer Nature Switzerland AG 2020
K. S. Mohamed, *Neuromorphic Computing and Beyond*,
https://doi.org/10.1007/978-3-030-37224-8